SAT® Math Prep Secrets
How to Achieve a Perfect Score on the SAT® Math Test

Cason Smart

First Edition 2019
Copyright © 2019 Richard Van Winkle
All rights reserved, including the right to reproduce
this book or portions thereof in any form whatsoever.
Cover Layout: Richard Van Winkle
Cover Design: Wolkenart - Marie-Katharina Wölk, www.wolkenart.com
Copy Editor: Bob Farrow, www.bfcpro.com
ISBN: 9781709129520
Printed by Amazon Distribution

Preface

When I began German engineering school in the fall of 2014, I had not passed a math class since I received a C in Algebra II, eight years prior. In the years following my lack of mathematical high school success, I managed to flunk Pre-Calculus twice before I finally gave up and joined the Army as a mechanic. It was during my time as a mechanic in the army that I decided that I wanted to study mechanical engineering. I was stationed in Germany and drove down to the local university and asked how I could study there. It would be possible, but I would have to learn math. My heart started racing at the realization, given my past success, or lack thereof, in mathematics. However, as I walked back to the car, I decided that I could learn how to do mathematics. This decision, this belief, that you can learn and master mathematics is the most important step on your journey to a perfect score on the math portion of the SAT test.

In the first semester at the German engineering school I found myself in a combined Calculus I & II course. The task seemed impossible and I was desperate for any kind of help in mathematics. The first week at school I saw a flier for free tutoring, and it was in this tutoring that I met Thomas. Within the first three weeks of tutoring with him he had shown me the simple, yet most powerful, secrets to success in mathematics. It was because of these secrets that I was able pass my first semester of the combined Calculus I & II with a B average. The next semester covered Calculus III and differential equations; I received no less than an A on that course. I took these same secrets with me over the next four years of college and received A's in every mathematical based engineering course on the way to my degree. Over the next pages I am going to share those same secrets with you. I am beyond confident that these secrets will propel you, not only to a perfect SAT score, but also to years of success at collegiate level.

Contents

Preface   iii

1   How to Use This Book   1

2   Essential Reference Guide   3
   2.1   Common Sets of Numbers 3
   2.2   Absolute Value 4
   2.3   Sums and Products 4
   2.4   Rules of Fractions 5
   2.5   Exponents and Roots 6

3   Heart of Algebra   9
   3.1   Absolute Value 9
   3.2   Linear Equations 15
   3.3   Linear Inequalities 20
   3.4   Systems of Linear Equations 23
   3.5   Systems of Linear Inequalities 30
   3.6   Relationship between Linear Equations and Lines in the Coordinate Plane 36

4   Passport to Advanced Math   40
   4.1   Systems of Equations 40
   4.2   Operations with Polynomials and Rewriting Expressions 44
   4.3   Quadratic Functions and Equations 48
   4.4   Solving Rational Equations 51
   4.5   Function Notation 53
   4.6   Relationships Between Algebraic and Graphical Representations of Functions 56
   4.7   Exponential Functions, Equations, Expressions and Radicals 61
   4.8   More Complex Problems in Context 68

5   Problem Solving and Data Analysis   73
   5.1   Ratio, Proportion, Units, and Percentage 73
   5.2   Interperting Relationships Presented in Scatterplots, Graphs, Tables, and Equations 80
   5.3   Understand and Analyze Data Presented in a Table, Bar Graph, Histogram, Line Graph or Other Display 95

6   Additional Topics in Math   106
   6.1   Geometry 106
   6.2   Trigonometry 114

7   Practice Test 1   **117**
   7.1   Practice Test 1 Solutions 142
   7.2   Practice Test 1 Explanations 143

8   Practice Test 2   **165**
   8.1   Practice Test 2 Solutions 189
   8.2   Practice Test 2 Explanations 190

9   Practice Test 3   **208**
   9.1   Practice Test 3 Solutions 234
   9.2   Practice Test 3 Explanations 235

Chapter 1

How to Use This Book

There are many ways to shear a sheep. But, at the end of the day all that matters is that you have the wool. There are many ways to get a perfect score on the SAT® test. But, at the end of the day all that matters is that you see an 800 on your results. This book is not here to teach you how to analyze the SAT test into its smallest detail. This book will show how simple it is to build the mathematical foundation necessary to crush the SAT math test.

Where other books might explain to you special techniques to you that have been set up just to beat the SAT® test, I asked myself the question: What is easier, memorizing a laundry list of special techniques with little to no logical foundations or reinforcing the mathematical foundation that you have been building since elementary school? I promise you, that with only a few rules that you will find in this book, some of which you already know very well, you can fill in the gaps in your mathematical foundation necessary to make the SAT® test feel like a coloring book.

All be it, the instructions in this book are only 30% of the battle. The first 70% of the battle can be won by simply changing your mindset. If you think you can't do math or that math is just something you can't understand. Then let me tell you the greatest secret of all: **You can do math**. Believe it, because you can. The second you start believing that you can do math, you will notice that everything will come easier to you.

When you throw yourself into this book, you might find yourself making many mistakes. Mistakes are good when you are learning math. In fact, if you don't make any mistakes then you are either ready to ace the SAT® or you are not trying hard enough. My general rule of thumb when learning a new mathematical concept is that you are allowed 1,000 mistakes before you give up. I have yet to reach 1,000 mistakes nor have I met a mathematical concept I can't conquer. That I attribute not to my being "smart" but because I believe that I can do math and I always allow my self to give up after 1,000 mistakes.

I have set this book up not with the special techniques taught by most SAT® prep books. Rather I have included problems including everything you might possibly see in the SAT test. Every problem is there with a solution and explanation section. I recommend you first try to solve the problem on your own, but if you cannot do that the first time you see the problem then go ahead and walk through the solution to the problem provided for you. When going over the solution, try to understand why you are allowed to make every mathematical decision on the path to the solution. When you understand the foundation behind the mathematical process, then you are solidifying your skill set in math.

I have chosen to omit explaining why all of the correct answers are incorrect. If you understand the reasoning behind how to get the correct answer, then I don't want to confuse you with an incorrect understand of mathematical processes which will lead you to the incorrect answer. Sometimes, on the path to the correct answer I will eliminate possible answers as being incorrect. In those cases, though, only because the path to the correct solution proves that it would be impossible for those solutions to be correct. On rare occasions I will explain why an answer is incorrect when the difference between the correct and incorrect solutions differ by a small often misunderstood detail.

There are two main portions to this book; chapters 3-6 go over sample problems from all the possible topics on the SAT test. Along with these problems is a very detailed path to the solution. Within the solution are built in the key foundational mathematical concepts necessary for success on the SAT test. Chapters 7-9 are full of three complete SAT® practices, one per chapter. After each test is a page with only the solutions followed by a section with explanations of how to solve each problem from the practice test. I recommend that you stick to the time and calculator constraints when using the practice tests. If you find yourself struggling, go ahead and work through the solutions of the questions you are having trouble with, always clarifying the mathematical reasoning behind the solution to yourself. Once you are done going over the questions you are struggling with, give yourself a break of one or two days and try the practice test again under the time and calculator constraints.

I wish you the best of luck while working with this book and that you never forget: You can do math!

Chapter 2

Essential Reference Guide

The following sections outline the most basic of rules of mathematics. If you continually refer back to these rules when solving problems, you will very soon be ready to score perfect on the Math portion of the SAT® test.

2.1 Common Sets of Numbers

A true understanding of mathematics can only be realized when you know its limitations. The following sets of numbers define where a math problem can start and where it can end. I do not expect you to study these extensively; however, I will refer back to them from time to time when a math problem or concept requires their definition.

Natural Numbers

The set of the natural numbers is denoted with the symbol $\mathbb{N}$. This set of numbers is commonly referred to as the counting numbers. It begins at either 0 or 1 and continues by adding 1 to the previously counted number. For example, 0, 1, 2, 3, 4, etc. until you reach infinity. Mathematically the set of the natural numbers is written: $\mathbb{N} = \{0, 1, 2, ...\}$

Integers

The set of integers is denoted with the symbol $\mathbb{Z}$. Like the natural numbers, the separation between numbers in the set of integers is with a difference of one. Unlike the natural number, integers extend over into negative numbers. Mathematically the set of the natural numbers is written: $\mathbb{Z} = \{..., -3, -2, -1, 0, 1, 2, 3, ...\}$

Rational Numbers

The set of rational number includes all terminating decimals and non-terminating repeating decimals. Rational numbers are denoted with the symbol $\mathbb{Q}$.

A terminating decimal is any decimal number where the numbers after the decimal do not go on forever. For example: the number $\frac{1}{4}$ is written in decimal form as 0.25. There is no more relevant information after the 5, therefore the decimal *terminates* at that location.

A non-terminating decimal is any decimal number where the numbers after the decimal repeat forever. For example: the number $\frac{1}{3}$ is written in decimal form as 0.333... . The threes in this situation will never terminate instead they will repeat forever.

Whether or not you write rational numbers as decimals or fractions is entirely your decision. I would recommend using fractional forms of numbers for two reasons:

1. You can write infinitely long numbers in a very condensed form. Refer to the number $\frac{1}{3}$ above.

2. You don't have to worry about making rounding errors.

Real Numbers

The set of real numbers is denoted with the symbol $\mathbb{R}$. It includes all the sets of numbers already discussed along with the irrational numbers. Famous numbers like $\pi = 3.14159265$ and $\sqrt{2} = 1.41421356$ are irrational because the numbers after the decimal will never repeat nor terminate.

Complex Numbers

A complex number, z, is a number with a real component and an imaginary component. The imaginary component of a complex number contains the imaginary number, which is commonly denoted as i or j. The imaginary number is the square root of negative 1, $\sqrt{-1}$. In its most general form the complex number looks as follows:

$$z = a + bi$$

Where the variable a is the real component and the product bi is the imaginary component of the complex number.

2.2   Absolute Value

The absolute value $|a|$ of a real number a is the distance between the number and zero.

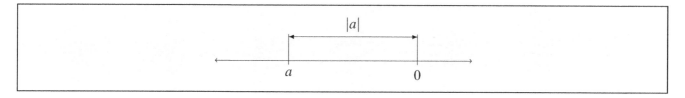

2.3   Sums and Products

Sums

A sum is defined as multiple numbers added or subtracted together. For example:

$$a \pm b \pm c \pm \ldots \pm x \pm y \pm z$$

Every element in the sum is referred to as a term. So that in the most general form a sum can be read as:

$$\text{term} \pm \text{term} \pm \text{term} \pm \ldots \pm \text{term} \pm \text{term} \pm \text{term}$$

Throughout the book when I refer to a term I will be talking about an element in a sum, as just described.

Products

A product is in its most basic form can be defined as numbers multiplied together. It looks as follows:

$$a \times b \times c \times \ldots \times x \times y \times z$$

The sign $\times$ is commonly used for multiplication; however, it is also common when you are multiplying a constant by a variable or a number by a variable that you leave out the multiplication sign all together. For example, in the product $3x$ or ab the $\times$ sign is dropped and multiplication is implied.

Another way to express multiplication is with the use of parenthesis. This is very common when multiplying a positive number with a negative number. For example, in the product $(-4)4$ the $\times$ sign is dropped, and multiplication is implied.

The individual elements of a products are referred to as different things depending on the context.

If you are multiplying a known value by a variable, for example in the case of $3x$, the 3 would be referred to as a factor. You would say x is being multiplied by a factor of 3.

If you are multiplying an unknown constant value by a variable, in the case of ax, the a would be referred to as a coefficient. In more professional circumstances where scientists might play around with coefficient values to get a desired outcome, a coefficient might also be referred to as a parameter.

2.4   Rules of Fractions

Basic Terminology

A fraction can be written as $\frac{a}{b}$. Where a is called the numerator and b is called the denominator. The denominator of a fraction can never equal 0.
When $|a| < |b|$ then the fraction is called a proper fraction. However, if $|b| < |a|$ the fraction is called an improper fraction.

Reciprocals

A very important property of fractions is that any integer number can be written as a fraction. For example, the number 2 is no different than the number $\frac{2}{1}$. To find the reciprocal of a fraction you need only flip the fraction over, i.e. the numerator becomes the denominator and vice versa. Given this description the reciprocal of the number $\frac{3}{2}$ is $\frac{2}{3}$. The reciprocal of the number 2 or $\frac{2}{1}$ is $\frac{1}{2}$.

$$\text{The reciprocal of} \quad \left\{ \begin{array}{c} a \\[2mm] \dfrac{a}{b} \end{array} \right. \text{is} \left. \begin{array}{c} \dfrac{1}{a} \\[2mm] \dfrac{b}{a} \end{array} \right\} \quad (\text{where } a \neq 0 \text{ and } b \neq 0)$$

Fraction Addition and Subtraction

The following method can always be used to add two fractions.

$$\frac{a}{b} + \frac{c}{d} = \frac{a \times d + c \times b}{b \times d}$$

The following method can always be used to add three fractions.

$$\frac{a}{b} + \frac{c}{d} + \frac{e}{f} = \frac{a \times d \times f + c \times b \times f + e \times b \times d}{b \times d \times f}$$

Subtraction of fractions can be accomplished using the same methods. To accomplish the subtraction, replace the addition symbol with a subtraction sign.

Fraction Multiplication

Multiply fractions by multiplying the respective numerators together and the respective denominators together.

$$\frac{a}{b} \times \frac{c}{d} = \frac{a \times c}{b \times d}$$

Fraction Division

To divide one fraction by another, multiply the dividend by the reciprocal of the divisor.

$$\frac{a}{b} \div \frac{c}{d} = \frac{a}{b} \times \frac{d}{c} = \frac{a \times d}{b \times c}$$

2.5 Exponents and Roots

Exponents and roots have a handful of very simple rules. To be able to fluently work with them you only need to understand basic arithmetic and fraction arithmetic.

The rules for exponents are as follows:

1. $a^m \times a^n = a^{m+n}$

2. $\dfrac{a^m}{a^n} = a^{m-n}$

3. $(a^m)^n = (a^n)^m = a^{m \times n}$

4. $a^n \times b^n = (a \times b)^n$

5. $\dfrac{a^n}{b^n} = \left(\dfrac{a}{b}\right)^n$

For each of the exponent rules above the following must be true:

- m and n are natural numbers greater than 0.

- a and b are real numbers.

- Anytime a or b appear in the denominator of a fraction it cannot be equal to 0.

The rules for dealing with roots are very similar to those involving exponents. Every rule involving roots builds from the first rule which is as follows:

$$\sqrt[n]{a} = a^{\frac{1}{n}}$$

The above rule is probably one of the most important tools a mathematician can have in his tool belt. The remaining rules for roots are as follows:

$$1. \quad \sqrt[n]{a^m} = (a^m)^{\frac{1}{n}} = a^{\frac{m}{n}} = (a^{\frac{1}{n}})^m = (\sqrt[n]{a})^m$$

$$2. \quad \sqrt[m]{\sqrt[n]{a}} = \sqrt[m]{a^{\frac{1}{n}}} = (a^{\frac{1}{n}})^{\frac{1}{m}} = a^{\frac{1}{m \times n}} = \sqrt[m \times n]{a}$$

$$3. \quad \sqrt[n]{a} \times \sqrt[n]{b} = (a^{\frac{1}{n}}) \times (b^{\frac{1}{n}}) = (a \times b)^{\frac{1}{n}} = \sqrt[n]{a \times b}$$

$$4. \quad \frac{\sqrt[n]{a}}{\sqrt[n]{b}} = \frac{a^{\frac{1}{n}}}{b^{\frac{1}{n}}} = \left(\frac{a}{b}\right)^{\frac{1}{n}} = \sqrt[n]{\frac{a}{b}}$$

For the rules above the following must be true:

- m and n must be natural numbers and greater than 0.

- a and b must be greater than or equal to 0.

- Anytime a or b appears in the denominator of a fraction it cannot be equal to 0.

Unit Conversion

The rule in algebra is that whatever you do to one side of the equation you must do to the other side. There are very few exceptions to this rule. One of which appears when you need to convert the units of a physical measurement.

The exceptions to the rule are as follows:

You may operate on only one side of an equation or only one term in an equation if:

- You are multiplying by 1

- You are adding 0

In the case of unit conversion, we are multiplying a single term by 1. I will show how this works with an example.

Unit Conversion Example

A farmer wants to irrigate his 150 acres of vineyards using drip irrigation. The company quoted him a price of 23 cents per square foot, what is the price in dollars per acre? There are 43,560 square feet in an acre.

Unit Conversion Solution

Here's the thing, every unit conversion ratio must equal 1. In this case we know that 43,560 square feet is 1 acre or written mathematically:

$$43,560 \text{ square feet} = 1 \text{ acre}$$

Let's convert the left side of this equation to a ratio by multiplying the entire equation by $\dfrac{1}{1 \text{ acre}}$.

$$\frac{1}{1 \text{ acre}}(43,560 \text{ square feet} = 1 \text{ acre})$$

$$\frac{43,560 \text{ square feet}}{1 \text{ acre}} = \frac{1 \text{ acre}}{1 \text{ acre}}$$

Notice how the right side of the equation reduces to 1 because the values in the numerator and denominator are now the same. Which means the equation becomes:

$$\frac{43,560 \text{ square feet}}{1 \text{ acre}} = 1$$

We just proved that a unit conversion ratio is equal to 1. Because of this we are allowed to multiply it with any term in an equation. So, in the question above when we are asked to convert the price from cents per square foot to dollars per acre. The equation before the conversion looks as follows:

$$\text{price per acre} = \frac{23 \text{ cents}}{1 \text{ square foot}}$$

Normally if we were to do anything to this equation, we would have to do it to both sides of the equation. However, because we are doing a unit conversion, multiplying by 1, we can perform this operation only to the right side of the equation.

$$\text{price per acre} = \frac{23 \text{ cents}}{1 \text{ square foot}} \left(\frac{43,560 \text{ square feet}}{1 \text{ acre}} \right)$$

Notice how the units of square feet are in the numerator and denominator of the fraction, so that they cancel each other out leaving only units of cents per acre. After the multiplication the price per acre becomes:

$$\text{price per acre} = \frac{1,001,880 \text{ cents}}{1 \text{ acre}}$$

To convert the price from cents to dollars we need to know the relationship that 100 cents is equal to 1 dollar. That is written mathematically as follows:

$$1 \text{ dollar} = 100 \text{ cents}$$

I recommend you go through and turn this into a ratio set equal to one as we did with the relationship between square feet and acres. Once you prove that this unit conversion ratio is equal to 1, it should not bother you when you multiply the ratio by the term on the right side of the price per acre equation.

$$\text{price per acre} = \frac{1,001,880 \text{ cents}}{1 \text{ acre}} \left(\frac{1 \text{ dollar}}{100 \text{ cents}} \right)$$

Notice how the cents cancel each other out in this equation leaving only dollars per acre as the units. After the division by 100 the final answer is:

$$\text{price per acre} = \frac{10,018.80 \text{ dolars}}{1 \text{ acre}}$$

Chapter 3

Heart of Algebra

3.1   Absolute Value

Problem 1

A typical lasagna has the second layer of noodles more than 3 centimeters (cm) and less than 5 cm from the top of the lasagna. Which of the following inequalities describes all of the possible heights d, in cm, below the top of the lasagna that are in the second noodle layer?

A) $|d+3| < 5$

B) $|d-3| < 5$

C) $|d+4| < 1$

D) $|d-4| < 1$

Solution and Explanation

We know that the second noodle layer in a typical lasagna is between 3 cm and 5 cm below the top of the lasagna. Given that information we can build a range for d.

$$3 < d < 5$$

In order to solve this problem, we need to find an absolute value inequality equivalent to the inequality above. To do that we need to first find the midpoint of the second noodle layer. The midpoint between two numbers on the number line is also the average of the two numbers. The equation for finding the average of two numbers is to first find the sum of the numbers and then to divide those numbers by 2. using the numbers from this problem the mathematical process for finding the average looks as follows:

$$\frac{3+5}{2} = \frac{8}{2} = 4$$

Now pay attention closely.

The distance between any possible height inside the lasagna d and the midpoint of the second layer of noodles is the *absolute value* of the *difference* between any possible height, d, and the midpoint of the second noodle layer height, 4.

The last sentence is written mathematically as follows:

$$|d-4|$$

9

Now we need to focus on how far from the midpoint of the second noodle layer we can travel be[...]
layer. Namely we can only travel 1 cm in either direction before we leave the layer. If we move 1.1[...]
midpoint we will either be at $d = 2.9$ cm of $d = 5.1$ cm which is outside of the second noodle layer we[...]
earlier as $3\text{cm} < d < 5\text{cm}$.

Let's follow this logically, if the distance from the midpoint of the second noodle layer is $|d - 4|$ and this distan[...]
cannot be greater than 1 cm, then the following mathematical statement is true:

$$|d - 4| < 1$$

The correct answer is D.

fore we exit that
cm from the
defined

$$y = |x - 2|$$
$$y = 3x + 6$$

)ve set of equations?

$$x - 2 = 3x + 6$$
$$+2 \qquad +2$$

$$x - 2 = 0$$

$$x = 3x + 8$$
$$-3x \quad -3x$$

$$|x - 2| = 0$$

$$\frac{-2x}{-2} = \frac{8}{-2} \qquad x = -4$$

C) (-4,-6)

$$x - 2 = 0$$
$$+2 \quad +2$$

D) (-4,6)

$$x = 2$$

Solution and Explanation

This problem looks to combine your skills with absolute values with your skills of solving sets of linear equations. The first thing you want to do with this problem is set up systems of equations without the absolute value. In order to do that you need to remember that the absolute value is the measure of a number's distance from zero and will always be greater than zero.

Start by looking for the value of x which will turn the value inside the absolute value to zero. Set up an equation to do that:

$$|x - 2| = 0$$

Here we can temporarily get rid of the absolute value symbols, because we are only looking for a value of x that will make the $x - 2$ expression zero.

$$x - 2 = 0$$

To solve for x add 2 to both sides of the above equation.

$$x - 2 + 2 = 0 + 2$$
$$x = 2$$

By solving for the value of x which brings the absolute value to zero, we can begin to develop ranges for x which will bring the value inside the absolute value to a negative and a positive number. For example, if you were to increase 2 to 2.001, then 2.001 - 2 = 0.001 and the value inside the absolute value is a positive number. In other words, for values greater than three there is no need for the absolute value sign, because the value will always be positive. Write down the first range of x for the absolute value equation as follows:

$$\text{For } x \geq 2 \qquad y = x - 2$$

Now if you were to decrease the number 2 to 1.999, then 1.99 - 2 = -0.01. Since the numbers under 3 will result in a negative number, you need to multiply them with a negative 1 in order to get them back to being positive. Given that, we can again drop the absolute value symbol and replace it with a convenient equation for values of x less than zero. Within this range the equation would look as follows:

$$\text{For } x \leq 2 \qquad y = -(x - 2)$$

11

Notice that in the equations for both ranges I use the less than or equal to, and the greater than or equal to, symbols. This is because zero is neither positive nor negative, it is simply zero, and both the equations hold true for $x = 2$.

Now what we have done here is actually created two sets of equations, one for each range of x whether it be greater than or less than 2. The two sets of equations look as follows:

For $x \geq 2$:

$$y = x - 2$$
$$y = 3x + 6$$

For $x \leq 2$:

$$y = -(x - 2)$$
$$y = 3x + 6$$

Let's start with the range for $x \geq 2$, the top system of equations. Both equations are already set equal to y, so we can set the right sides of both equations equal to each other and solve for x.

$$x - 2 = 3x + 6$$

To solve the above equation for x you should move all of the non-x terms onto the right side of the equation and all of the x terms onto the left side of the equation. Start by moving the non-x terms to the right side of the equation; to accomplish that add 2 to both side of the equation.

$$x - 2 + 2 = 3x + 6 + 2$$
$$x = 3x + 8$$

Now to move all the x terms to the left, subtract $3x$ from both sides of the equation.

$$x - 3x = 3x - 3x + 8$$
$$-2x = 8$$

The final step to solve for x is to divide the entire equation by -2. in other words multiply the entire equation be $-\frac{1}{2}$.

$$-\frac{1}{2}(-2x = 8)$$
$$-\left(\frac{-2}{2}x\right) = -\frac{8}{2}$$
$$x = -4$$

Although we were able to find a solution to the equation. This solution is not valid, because we are looking at the range for $x \geq 2$ and negative 4 does not fall within this range. So, we have analyzed one system of equations but still have not come to a solution. Let's take a look at the other system of equations to see if the solution is there.

For $x \leq 2$ it is again the case that both equations are already set equal to y, so again we can set both right sides from the system of equations equal to each other.

12

$$-(x-3) = 3x+6$$

The first thing that jumps out from this equation is the negative sign on the left side. Begin by distributing the negative sign into the terms on the left side of the equation.

$$-x+3 = 3x+6$$

Here again you should move all the non-x terms to the right side of the equation and all the x terms onto the left side of the equation. Begin with the non-x terms by subtracting 3 from both sides of the equation.

$$-x+3-3 = 3x+6-3$$
$$-x = 3x+3$$

To bring the x terms to the left side subtract $3x$ from both sides of the equation.

$$-x-3x = 3x-3x+3$$
$$-4x = 3$$

The final step to solve for x is to divide the entire equation by negative 4. In other words, multiply the entire equation by $-\frac{1}{4}$.

$$-\frac{1}{4}(-4x = 3)$$
$$-\left(\frac{-4}{4}x\right) = -\frac{3}{4}$$
$$x = -\frac{3}{4}$$

Notice how this time the value for x falls within the range of $x \leq 2$. Because of that this is a valid solution. We can plug this value for x into either one of the equations from the systems of equations for the range $x \leq 2$. I would suggest using the top equation from the two because the x term there is only multiplied by a negative 1 instead of a 3.

$$y = -(x-2) = -\left(-\frac{3}{4}-2\right)$$

The first thing to do to make this equation easier to solve is to distribute the negative into the terms inside the parenthesis.

$$y = \frac{3}{4}+2$$

Now the problem is that we need to add a fraction with a whole number. In order to add these two numbers, they both need to have the same number in the denominator. Usually when we perform an operation on a mathematical equation, we apply it to the entire equation. Here is an exception, if you multiply a term in the equation by 1, you do not need to apply that operation to the entire equation. We are going to multiply the 2 in this equation by $\frac{4}{4}$ or 1. Again, because this is multiplication by 1 it does not need to be applied to the entire equation.

$$y = \frac{3}{4}+2\frac{4}{4} = \frac{3}{4}+\frac{8}{4}$$

To add the two fractions simply add the numbers from each numerator with one another.

$$y = \frac{3}{4} + \frac{8}{4} = \frac{3+8}{4} = \frac{11}{4}$$
$$y = \frac{11}{4}$$

The correct answer is B.

3.2 Linear Equations

Problem 1

If $\frac{1}{5}f + \frac{1}{4}g = 2$, what is the value of $4f + 5g$?

Solution and Explanation

Remember that whatever you do to one side of an equation, you *must* do to the other. Whenever you see an equation full of fractions immediately ask yourself, "how can I get rid of these fractions?" Here's how you do it; look at the denominator of both of these fractions. There is a 5 and a 4, to get rid of these numbers in the denominator we need to put a number in the numerator that is both divisible by 5 and 4. Instead of sitting there and trying to figure out what the number is, just multiply 5×4 to get the number 20. Now we take the entire equation and multiply it by the number 20.

$$20 \times \left(\frac{1}{5}f + \frac{1}{4}g = 2 \right)$$

If what you see above scares you, I would like you to walk through the next few steps three or four times. Enough times until you become confident that you too can pull off this professional mathematical trick.

The next step is to distribute the 20 over the entire equation.

$$20 \times \frac{1}{5}f + 20 \times \frac{1}{4}g = 20 \times 2$$
$$\frac{20}{5}f + \frac{20}{4}g = 40$$

You have just successfully multiplied an entire equation by 20. Now let's get rid of the fractions by dividing 20 by 5 and 4 in this next step.

$$4f + 5g = 40$$

Look closely at the above equation. It is exactly that what the problem was asking for, namely that $4f + 5g = 40$.

The correct answer is 40.

15

Problem 2

Solve this equation in terms of the variable x.

$$3(-x+5) + (3x-4) = 2(4x-3)$$

Solution and Explanation

The SAT® committee wants to test a multitude of different skill sets. Mathematically the truth is this: If you hold fast to only a few mathematical pillars, you can conquer any problem. In this problem they have given you a cluster of an equation. However, note that behind this cluster there is only *one* unknown, namely the variable x, and only *one* equation; now you have the essential knowledge base required to solve this problem.

The next steps we are going to perform mathematically will distribute any factors and rearrange the sums so that only numerical values are on the right side and variables with their numerical factors are on the left side.

$$-3x + 15 + 3x - 4 = 8x - 6$$
$$-3x + 3x - 8x = -15 + 4 - 6$$
$$-8x = -21 + 4$$
$$-8x = -17$$

When both sides of the equation have a negative sign in front of them the best thing to do is to multiply the entire equation by -1.

$$-1 \times (-8x = -17)$$
$$8x = 17$$

Now all that is left to accomplish is to get the variable x all by its lonesome. to do that divide the entire equation by 8, i.e. multiply the equation by $\frac{1}{8}$.

$$\frac{1}{8} \times (8x = 17)$$
$$\frac{8}{8}x = \frac{17}{8}$$
$$x = \frac{17}{8}$$

The solution to this problem is $x = \frac{17}{8}$. Often in your college education it will be the case that the solution is in the form of a fraction. Do not be scared by this fact, rather embrace it as a more exact solution.

The correct answer is $\dfrac{17}{8}$

Problem 3

Tony received stones from his grandmother. For the next week Tony went down to the creek every day to collect more stones. The number of stones, S, Tony collected D days after his grandmother gave him the stones can be estimated by the equation $S = 30 + 12D$. What is the best interpretation of the number 12 in this equation?

A) The number of stones Tony's grandmother gave to him.

B) The total number of stones Tony had after a week of collecting stones.

C) The number of stones Tony collected each day that he went to the creek.

D) The number of stones Tony had at the end of the week.

Solution and Explanation

The solution A) is represented in the equation by the number 30. If we set D equal to 0, which means the day before Tony started collecting stones, Tony would have had 30 stones. Setting the variable equal to 0 gives the initial value, which in terms of the problem represents the number of stones Tony received from his grandmother.

In order for B or D to be correct we would need to insert 7 in the place of the variable D. This question is however only asking what 12 represents, not how many stones Tony has after a certain number of days.

Each time D increases, for every day that passes, 12 more stones are added to the total number of stones collected, S. Therefore 12 represents the number of stones Tony collected each day.

The correct answer is C.

Problem 4

$$\frac{4}{7} + 21s = 6\left(\frac{1}{4} - s\right)$$

What is the solution to the equation above?

Solution and Explanation

The variable we need to solve for in this equation is s. When I first look at this equation there are two things that bother me from the get-go. Those are the terms inside of the parenthesis and the fractions. The easiest thing to get rid of are those in parenthesis, so we should get rid of those first. We can do that by distributing the 6 inside of the parenthesis. Considering the entire equation that looks as follows:

$$\frac{4}{7} + 21s = \frac{6}{4} - 6s$$

You can make life easier for yourself by factoring out and simplifying the fraction $\frac{6}{4}$. That looks like this:

$$\frac{6}{4} = \frac{3 \times 2}{2 \times 2} = \frac{3}{2}$$

Now our equation looks as follows:

$$\frac{4}{7} + 21s = \frac{3}{2} - 6s$$

The next step is to get rid of the fraction. Because there is a 7 in one denominator and a 2 in the other denominator, we can multiply the entire equation once by 7 and then again by 2 and the fractions will disappear. First let's multiply everything by 7.

$$7\left(\frac{4}{7} + 21s = \frac{3}{2} - 6s\right)$$

$$\frac{7 \times 4}{7} + 7 \times 21s = \frac{7 \times 3}{2} - 7 \times 6s$$

$$4 + 147s = \frac{21}{2} - 42s$$

Now we can go ahead and multiply the entire equation by 2.

$$2\left(4 + 147s = \frac{21}{2} - 42s\right)$$

$$2 \times 4 + 2 \times 147s = \frac{2 \times 21}{2} - 2 \times 42s$$

$$8 + 294s = 21 - 84s$$

After getting rid of the parenthesis and fractions the equation looks a lot more manageable. The next steps are to get all the like terms on their own sides of the equation. First, we will move all the terms containing s to the left side of the equation and then combine them. We can do that by adding $84s$ to both sides of the equation.

$$8 + 294s + 84s = 21 - 84s + 84s$$

$$8 + 378s = 21$$

Now bring the 8 over to the other side of the equation by subtracting 8 from both sides of the equation.

$$8 - 8 + 378s = 21 - 8$$
$$378s = 13$$

The final step in solving for s is dividing the entire equation by 378 in other words multiplying the entire equation by $\frac{1}{378}$.

$$\frac{1}{378}(378s = 13)$$
$$\frac{378}{378}s = \frac{13}{378}$$
$$s = \frac{13}{378}$$

The correct answer is $\frac{13}{378}$.

3.3   Linear Inequalities

Problem 1

The recommended protein intake for a 36-year-old is 46 grams. One steak contains 8 grams of protein and an egg contains 5 grams of protein. Which of the following inequalities represents the number of steaks s and eggs e a 36-year-old needs to eat in order to meet or fall short of the recommended daily intake of protein?

A) $8s + 5e < 46$

B) $\frac{8}{s} + \frac{5}{e} \leq 46$

C) $\frac{8}{s} + \frac{5}{e} < 46$

D) $8s + 5e \leq 46$

Solution and Explanation

The first thing to do with this problem is to clearly define what is being asked. Reread the question and consider that before reading the next sentence. Here we need to know how many steaks and eggs, in combination, the man must eat in order to receive 46 or less grams of protein.

With a clear definition of what is required you can see what components of the question you need to consider. In this case it is any number of steaks s and eggs e needed to get to 46 grams of protein. If the man eats one steak, he will receive 8 grams of protein, if he eats two steaks, he will receive 16 grams of protein. Because the number of grams per steak eaten are added together, we are dealing with multiplication. Knowing this you can already eliminate two of the possible choices, the two where s and e are placed in the denominators of the fractions.

The only thing left to consider is the inequality sign in the equation. If the man were to eat exactly 46 grams of steak, he would meet the daily requirement. That scenario is represented mathematically with an equal sign. Since there are no equal signs in the possible answers, the correct sign will be less than or equal to. You can also tell by the wording "fall short of" that a less than sign will need to be a part of the answer.

The correct answer is D.

Problem 2

If $-\dfrac{17}{3} < -7r + 4 < -\dfrac{12}{5}$ what is one possible value of $28r - 16$?

Solution and Explanation

This problem requires a trained mathematical eye. At first it might seem overwhelming, but as soon as you notice that you can get $28r - 16$ by multiplying $-7r + 4$ by negative 4, you are in business. This however is really the only difficult part of the problem. When studying for the SAT or math in general, it helps to practice rare scenarios multiple times when you come across them, so that you develop what my Calculus teacher called "mathematical maturity." This problem is one of those that I would recommend practicing a few times, and then skipping a couple of days in between. By doing this you will begin to notice when one equation is a multiple of another.

Back to the problem at hand, we need to replace the $-7r + 4$ in this equation with $28r - 16$. This means we need to multiply the entire inequality by negative four. This brings about a very particular point about inequalities, which is:

- Whenever you multiply or divide an inequality by a negative number, you need to flip the inequality sign around.

Let's do the math, keeping the above fact in mind.

$$-4\left(-\frac{17}{3} < -7r + 4 < -\frac{12}{5} \right)$$

$$\frac{(-17) \times (-4)}{3} > -4(-7r + 4) > \frac{(-12)(-4)}{5}$$

Notice above how the negative 4 distributes itself throughout the entire inequality. Also notice that since we multiplied the entire inequality by negative 4 the inequality signs are now flipped around. In the next step we will perform the multiplication in the fraction on the outer sides of the inequality and the distribution on the inside of the inequality.

$$\frac{68}{3} > 28r - 16 > \frac{36}{5}$$

For the answer to this problem you can write in any number **less than** $\dfrac{68}{3}$ and **greater than** $\dfrac{36}{5}$.

Problem 3

A manager for a manufacturing firm estimates the cost c, in Euros of producing n products is $15n + 525$. The firm sells each product for 22 Euros. The firm makes a profit when the total income for selling a quantity of products is greater than the total cost of producing that quantity of products. Which of the following inequalities gives all possible values of n for which the manager estimates that the company will make a profit?

A) $n < 75$

B) $n > 75$

C) $n > 33$

D) $n < 33$

Solution and Explanation

This problem really boils down to being able to set up and solve an inequality. We know from the question that the total income from selling n products has to be greater than the cost of producing n products. The cost of producing n items is written in the questions as $15n + 525$. The total income for selling n products is equal to the price per product, 22 Euros, multiplied by the number of products sold, n. Knowing these two quantities we can set up the inequality.

$$22n > 15n + 525$$

Solving an inequality is very similar to solving an equation; however, you need to watch out for multiplication or division by a negative number. In which case you will need to flip the inequality sign around. For the inequality above we are first going to move all the terms with an n onto the left side of the inequality, leaving all non-n terms on the right side of the inequality. We can accomplish this by subtracting $15n$ from both sides.

$$22n - 15n > 15n - 15n + 525$$
$$7n > 525$$

Now in order to get n all by itself on the left side of the inequality we need to divide the inequality by 7, in other words multiply the equation by $\frac{1}{7}$.

$$\frac{1}{7}(7n > 525)$$
$$\frac{7}{7}n > \frac{525}{7}$$
$$n > 75$$

The correct answer is B.

3.4 Systems of Linear Equations

Problem 1

$$7x + 2y = 8 + 3x$$
$$5x + 2y = 9 - y$$

Based on the system of equations above, what is the value of the product xy?

Solution and Explanation

In this problem we are looking for the values of two variables, x and y. In either equation the variables x and y appear on both sides of the equation. The writers of the test do this on purpose because they want to see if you have mastered the most principal of algebraic concepts, which is:

- Whatever you do to one side of an equation, you *must* do to the other.

The first things we need to do to these equations are to get all the variables to one side, in this case the left side, and all the constants to the other side. We can accomplish this by subtracting $3x$ from both sides of the top equation and adding y to the bottom equation.

$$7x + 2y - 3x = 8 + 3x - 3x$$
$$5x + 2y + y = 9 - y + y$$

You can see above that the $3x - 3x$ on the right side of the top equation will be equal to zero, leaving the like terms $7x - 3x$ on the left side of the top equation to fall to $4x$. For the bottom equation on the right side, $-y + y$ turns out to be zero, while on the left side $2y + y$ will become $3y$. Which leaves us with the following set of equations:

$$4x + 2y = 8$$
$$5x + 3y = 9$$

At this point the path forward can be tricky if you are new at this. I am going to explain the step-by-step method to you. Then I will solve the problem with you while explaining each step. Then I ask you to try and solve the problem yourself one time each day without looking at the solution, while at the same time being able to independently explain the reasoning behind each step that you make. The steps that you need to make to solve this problem are as follows:

1. Solve one of the equations for a single variable.

2. Take the "solved for a single variable equation" and plug that variable into the other equation. At this point you should have only one variable within one equation.

3. Solve "the other equation" in terms of its variable. Now you have a numerical solution for one of the variables.

4. Take the numerical solution and plug it into the "solved for a single variable equation."

5. Solve for the remaining variable. You should now have a numerical solution for both variables.

As per the step-by-step method the first thing we need to do is solve one of the equations for a single variable. With practice you will notice that the top equation, when multiplied by $\frac{1}{2}$ will reduce the factor attached to the y down to 1. Leaving the variable y all by itself to be solved. Let's work exclusively with the first equation and see how that looks mathematically.

$$\frac{1}{2} \times (4x + 2y = 8)$$
$$\frac{4}{2}x + \frac{2}{2}y = \frac{8}{2}$$
$$2x + y = 4$$

Now rearrange the equation so that y can stand alone. To do that subtract $2x$ from both sides of the equation, refer to the steps used when we first simplified the set of equations.

$$2x + y - 2x = 4 - 2x$$
$$y = 4 - 2x$$

You have just solved the top equation for y in terms of x. Take this value for y and plug it into the bottom equation, this will leave you with only one equation containing the single variable x.

$$5x + 3y = 9 \qquad \text{with } y = 4 - 2x \text{ becomes}$$
$$5x + 3 \times (4 - 2x) = 9$$

To get rid of the term inside of parenthesis multiply the three across it.

$$5x + 12 - 6x = 9$$

Before solving for x move all the constant terms to the right side of the equation and all the variable terms to the left side. Once that is accomplished combine the like terms.

$$5x - 6x = 9 - 12$$
$$-x = -3$$

The problem with this equation is that everything is negative. We can clear that up very easily by multiplying the entire equation be -1.

$$-1 \times (-x = -3)$$
$$x = 3$$

Now that we know the value of x take that number and plug it into the equation we solved for y in terms of x. That was the equation $y = 4 - 2x$, after plugging that in we get the following solution for y:

$$y = 4 - 2 \times 3$$
$$y = 4 - 6$$
$$y = -2$$

The original question asked us to find the product of x and y. To do this just multiply to the solutions you have for x and y.

$$x \times y = 3 \times -2 = -6$$

The correct answer is -6.

Problem 2

Jane bought a pair of shoes and a hairband at a department store. The total price before taxes was $52.00. There was no sales tax on the hairband and a 14% sales tax on the pair of shoes. The total Jane paid, including the sales tax, was $54.64. What was the price, in dollars, of the hairband?

Solution and Explanation

At first glance this seems like a lot of information. Let's comb through it to see if we can define the variable and set up an equation or system of equations. The two variables in this equation are the shoes S and the hat H. There are two total prices in the description, which gives away that we will be dealing with a system of equations.
The first thing we know is that the hat and shoes, without tax, will add up to $52.00. Let's jot down that equation really quick.

$$S + H = 52$$

The other equation is the total of the items with the sales tax applied only to the pair of shoes. The sales tax of 14% can be written more practically as 1.14. Create a product with 1.14 and the variable as S, add that to the price of the untaxed hat and set it equal to the total with tax. The equation will look as follows:

$$1.14 \times S + H = 54.64$$

Now that we know both equations, we can write them down as a system of equations, I will only do this for the explanation. In the test I do not recommend writing down the equations a second time once you have already written them.

$$S + H = 52$$
$$1.14 \times S + H = 54.64$$

The question is asking only for the price of the hat. So, the best thing to do is eliminate the variable S as soon as possible. Solving the bottom equation for S will result in a fraction with a sum in the numerator; try it out for practice. But, in this case let's solve the top equation for S which will only result in a sum. To solve for S in the top equation subtract H from both sides.

$$S + H - H = 52 - H$$
$$S = 52 - H$$

Plug that solution into the bottom equation, which will result in the following:

$$1.14 \times (52 - H) + H = 54.64$$

Now we need to go through the algebraic steps to solve for the variable H. Let's solve this equation for H, follow along with the algebra. First, we are going to distribute the 1.14 in the above equation into the parenthesis.

$$59.28 - 1.14 \times H + H = 54.64$$

Subtract 59.28 from both sides and $1.14 \times H$ from the single H.

$$59.28 - 59.28 - 0.14 \times H = 54.64 - 59.28$$
$$-0.14 \times H = -4.64$$

This equation has negative terms on both sides. We could multiply the entire equation by negative one to get rid of the negative signs. However, in this case we can kill two birds with one stone by dividing the entire equation by negative 0.14, i.e. multiplying by $-\frac{1}{0.14}$.

$$-\frac{1}{0.14} \times (-0.14 \times H = -4.64)$$

$$\frac{0.14 \times H}{0.14} = \frac{4.64}{0.14}$$

$$H = 33.14$$

The price of the hat is $33.14.

Problem 3

A local grocery store sells dark chocolate bars for \$2.35 and milk chocolate bars for \$1.95. During a three-day period, a total of 159 dark and milk chocolate bars were purchased, and the total income was \$338.85. Solving which of the following systems of equations yields the number of dark chocolate bars, d, and the number of milk chocolate bars, m, that were purchased during the three-day period.

A) $$d + m = 338.85$$
$$2.35d + 1.95m = 159$$

B) $$d + m = 159$$
$$2.35d + 1.95m = \frac{338.85}{2}$$

C) $$d + m = 159$$
$$2.35d + 1.95m = 338.85$$

D) $$d + m = 159$$
$$2.35d + 1.95m = 338.85 \times 2$$

Solution and Explanation

This question requires us to build a system of linear equations. This is clear because all of the answers appear as systems of linear equations. But, another thing to notice here is that the question is asking us to find values for two unknown variables. One of the most important rules in math is that you need the same number of equations as unknown variables in order to solve for the unknowns.

We know from the problem that the total number of dark chocolate bars, d, and the total number of milk chocolate bars, m, sold were 159. Which means if we create a sum out of the two products sold, and set it equal to 159, we will have a valid equation.

$$d + m = 159$$

Looking at this equation we can already see that the choice A is incorrect.

The fastest way to solve this problem is to notice that the second equation for every answer has the exact same left side. The left side in those equations is multiplying the price of the respective chocolate bar by the number of chocolate bars sold. Performing this multiplication and then adding both types of chocolate bars sold will result in the total income from selling those chocolate bars. The problem states that the total income was \$338.85. This is on the right side of equation for choice C. Therefore, choices B and D are both incorrect as well.

The correct answer is C.

Problem 4

$$2q + 4r = 64$$
$$cq + dr = 16$$

In the system of equations above, c and d are both constants. If the system of equations has an infinite number of solutions, what is the value of cd?

Solution and Explanation

A system of equation has infinitely many solutions when all the equations are exactly the same. In order for you to proceed, you need to figure out how to make the bottom equation the same as the top. Notice that the right sides of both equations are both whole numbers. Since that is the case you can divide the right side of the top equation by the right side of the bottom equation. That will give away what you need to multiply the bottom equation by, in order to match the right sides of both equations. Once the right sides of both equations match up, then the coefficients c and d must match the coefficients of the $q-$ and r-terms.

Begin by performing the division of 64 by 16:

$$\frac{64}{16} = 4$$

The 4 is going to be your multiplier to get the bottom equation set equal to the top equation. Now go ahead and multiply the entire bottom equation by 4. That will look as follows:

$$4(cq + dr = 16)$$
$$4cq + 4dr = (4)16$$
$$4cq + 4dr = 64$$

The next step is to compare the coefficients for the q and r-term. Start with the q term and set up the following equation:

$$4c = 2$$

To solve the equation above all you need to do is divide the entire equation by 4. In other words multiply the entire equation be $\frac{1}{4}$.

$$\frac{1}{4}(4c = 2)$$
$$\frac{4}{4}c = \frac{2}{4}$$
$$c = \frac{1}{2}$$

Follow the same process to solve for d. Compare the coefficients of the r-terms. The equation will look as follows:

$$4d = 4$$

To solve the above equation, divide the entire equation by 4. In other words multiply the entire equation by $\frac{1}{4}$.

$$\frac{1}{4}(4d = 4)$$

$$\frac{4}{4}d = \frac{4}{4}$$

$$d = 1$$

After you have solved for both c and d the final step is to multiply them together to find the solution for cd.

$$cd = \frac{1}{2}(1) = \frac{1}{2}$$

The correct answer is $\frac{1}{2}$.

3.5 Systems of Linear Inequalities

Problem 1

Cody swims at 4 miles per hour and skateboards at 9 miles per hour. His goal is to swim and skateboard for a total of at least 17 miles in less than 3 hours. If Cody swims s and skateboards k miles, which of the following sets of inequalities represent Cody's goal.

A) $\dfrac{s}{4} + \dfrac{k}{9} \geq 3$

$s + k < 17$

B) $4s + 9k \geq 17$

$s + k < 3$

C) $\dfrac{s}{4} + \dfrac{k}{9} < 3$

$s + k \geq 17$

D) $4s + 9k < 3$

$s + k \geq 17$

Solution and Explanation

Although this question is not asking you to create a system of equations. Notice that because there are two desired outcomes, there are also two equations. The easiest equation of the two to figure out in this equation is that the sum of the miles swam and the miles on the skateboard need to be at the very least 17 miles. Which means the total distance traveled can either be exactly 17 miles, or more than 17 miles, resulting in the following equation:

$$s + k \geq 17$$

Seeing this equation, we can already eliminate the options A and B. The second part of the equation requires a very basic knowledge of physics. In that, you will need to have a basic understanding of how to use physical units to your advantage and an understanding of the relationship between speed, distance and time. For the understanding of using units to your advantage we will look at the relationship between speed, distance and time.

$$speed = \frac{distance}{time}$$

Looking at the above equation in term of its units tells an interesting story. Distance in this problem is measured in terms of miles. In general distance can also be measured in feet, inches, etc. Time in this equation is measured in hours. In general, time can also be measured in seconds, days, etc. When a distance is travelled over time you are dealing with speed. In this case the speed is measured in miles per hour. Pay attention that speed may also be expressed as inches per second or feet per hour. Either way you need to be able to notice that you are dealing with a distance over time and therefore a speed. Now, we will look at the above equation again in terms of the units of speed, distance and time.

$$\frac{\text{miles}}{\text{hour}} = \frac{\text{miles}}{\text{hour}}$$

The unit son the left side of the above equation are the same as the units on the right side of the equation. If you are unfamiliar with working with units, note here that an equation not only balances numerical values on either side of an equal sign, but also balances the units of an equation.

Back to the problem, we know that in options C and D there must be the number 3 on the right side of the equation. Looking at the problem we also know that 3 stands for 3 hours. So, the units on the right side of the equation are in hours. Which means the units on the left side of the inequality must also be in hours.

If we look at option D, it multiplies 4 (the speed) by s (the distance) and 9 by k. Let's look at the upper equation only in terms of its units to see if it checks out.

$$\frac{\text{miles}}{\text{hour}} \times \text{miles} + \frac{\text{miles}}{\text{hour}} \times \text{miles} < \text{hour}$$

$$\frac{\text{miles}^2}{\text{hour}} + \frac{\text{miles}^2}{\text{hour}} < \text{hour}$$

Looking at the units of the answer D we see that the units on the left side of the inequality are not the same as those on the right side. This is a major red flag. Because the units on both sides of the equation do not match, answer D cannot be correct.

This means that the answer C is the correct answer. Let's look at the units for this option to see if our theory holds.

$$\frac{\text{miles}}{\frac{\text{miles}}{\text{hour}}} + \frac{\text{miles}}{\frac{\text{miles}}{\text{hour}}} < \text{hour}$$

A fraction in the denominator of a fraction means fraction division. An easier way to think about it is that you are multiplying the numerator by the inverse of the fraction in the denominator. Remember that any number can be written in a fraction just by placing the number in the numerator and the number 1 in the denominator.

$$\frac{\text{miles}}{1} \times \frac{\text{hour}}{\text{miles}} + \frac{\text{miles}}{1} \times \frac{\text{hour}}{\text{miles}} < \text{hour}$$

Notice here that the unit of miles cancels out in each term in the sum on the left side of the equation, leaving only hours on the left and right side of the equation.

$$\text{hour} + \text{hour} < \text{hour}$$

Because the units on the left and right side of the equation are the same and we have eliminated all other possible solutions as they are incorrect.

The correct answer is C.

Problem 2

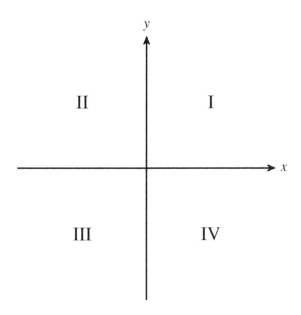

In which quadrant is there no solution for the following system of linear inequalities.

$$y \leq 2x - 3$$
$$y \geq 3x - 5$$

A) I

B) II

C) III

D) IV

Solution and Explanation

The solution to an inequality exists everywhere on a plane where the xy-coordinates, when put into the inequality, fulfill that inequality.

The most straight-forward way to solve systems of linear inequalities is to graph them and look at how the solutions overlap on the xy-plane. Let's begin by graphing the equality $y = 2x - 3$

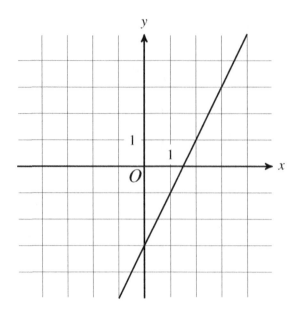

Given the line above you can now choose a coordinate from either side of the line and plug it into the equation. If the inequality holds true, then all of the points on that side of the line will satisfy the inequality. If the inequality does not hold true when the coordinates are plugged in, then all the solutions to the inequality are on the other side of the line.

To make life really easy on yourself choose the point (0,0) and plug it into the inequality.

$$y \leq 2x - 3$$
$$0 \leq 2(0) - 3$$
$$0 \leq -3$$

You can see that 0 is not less than -3, so this point does not satisfy the inequality. In other words, all the points on the same side of the line as (0,0) will not be solutions to this inequality. Rather, all the points on the side of the line not containing (0,0) will be solutions to the inequality and can be shaded in lightly with your pencil. Use the graph below for an example.

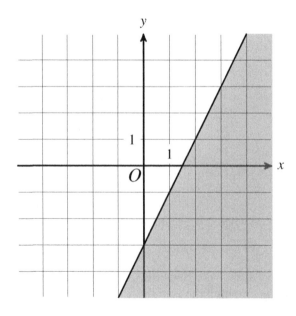

33

Here you can see that there is only one quadrant in which no solution exists. That is quadrant II. In which case B is the correct answer. On the SAT text I would answer the question at this point and move on.

For your further learning of how to graphically solve for systems of linear equations I will continue with the problem so that you can see and more importantly understand how systems of linear inequalities can be solved. Using the same method that you used for the upper equation graph the equation of the lower inequality, $y \geq 3x - 5$, to see where the line lies.

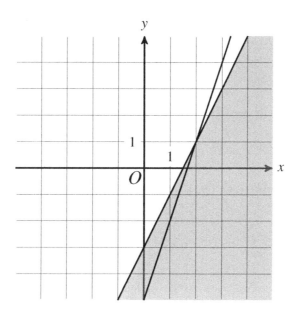

Here again the line does not run through the point (0,0) so it is a great point to use to test where the solutions for this inequality are.

$$y \geq 3x - 5$$
$$0 \geq 3(0) - 5$$
$$0 \geq -5$$

Since the above inequality holds true then all the points on the side of the line where the point (0,0) lies are solutions to the inequality. Which means that you can shade the region to the left of the line. Pay special attention to the region where the two solutions overlap. That region is the solution to the system of inequalities. I will shade that region in as a darker gray so that you can clearly see where the solution to the system of inequalities is.

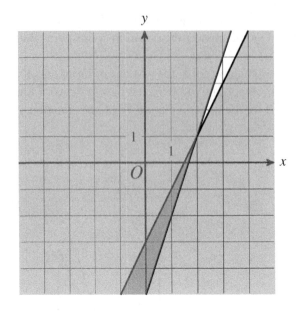

Here you can see that the solution to the system of linear inequalities does not lie in the second quadrant.

The correct answer is B.

3.6　Relationship between Linear Equations and Lines in the Coordinate Plane

Problem 1

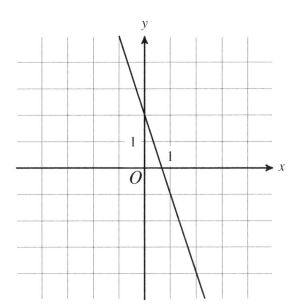

The graph of line s is shown above in the xy-plane. Which of the following equations is an equation of a line that is perpendicular to line s?

A) $y = -3x + 2$

B) $y = -\dfrac{1}{3}x - 2$

C) $y = \dfrac{1}{3}x - 3$

D) $y = 3x + 3$

Solution and Explanation

To find the slope of a line perpendicular to another line the product of the slopes needs to equal -1. To find the slope of a line pick out two points and build the ratio of the change in the y-direction over the change in the x-direction, known more commonly as the rise over run. For this problem, two easy points are (0,2) and (1,-1). The slope of the line in the above problem is:

$$\frac{\text{rise}}{\text{run}} = \frac{2 - (-1)}{0 - 1} = \frac{2 + 1}{0 - 1} = \frac{3}{-1} = -3$$

Remember that when you subtract a negative number it becomes addition.

Now build an equation with an unknown variable q to represent the slope of a line perpendicular to this one. That equation will be a product of the slope of the line in the graph above, with q set equal to negative one.

$$-3q = -1$$

To solve this equation for q divide the entire equation by negative 3, or multiply by $-\dfrac{1}{3}$. Keep in mind that a negative number divided by a negative number is a positive number.

$$-\frac{1}{3}(-3q = -1)$$

$$\frac{3q}{3} = \frac{1}{3}$$

$$q = \frac{1}{3}$$

The slope of a line perpendicular to the line shown in this problem is $\frac{1}{3}$. The only option above with that slope is option C, so that is the correct answer. Notice that wherever this line lies on the graph when it crosses the line shown, it will be perpendicular to it so the y-intercept is irrelevant.

Problem 2

$$-12r + bq = 4$$
$$3r - 5q - a = 2$$

In the system of equations above, a and b are constants. If the system has an infinite number of solutions, what is the value of a?

Solution and Explanation

For a system of linear equations to have infinitely many solutions the lines from the two equations need to be identical. Which in turn means the equations for the two lines need to be identical.

The two equations above are not identical. However, given that a and b are unknown constants we can solve for the values they would need to take on in order to make the equations identical. The first step is to set up the equations into similar forms. This can be accomplished by leaving all r and q terms on the left sides of the equations and bringing all the constant terms to the right sides of the equations. For these equations the only term that needs to be moved around is the constant a in the lower equation. Let's move it by adding a to both sides of the lower equation.

$$3r - 5q - a + a = 2 + a$$
$$3r - 5q = 2 + a$$

Now that both equations are in similar forms, we need to make them equivalent. One way to make sure they are equivalent is to make the r terms equivalent. This can be accomplished by either dividing the top equation by -4 or by multiplying the bottom equation be -4. In this case I will go with the multiplication to avoiding building fractions into the equations. Let's multiply the bottom equation by -4, that will look as follows:

$$-4(3r - 5q = 2 + a)$$
$$-4 \times 3r - (-4) \times 5q = -4 \times 2 + (-4)a$$
$$-12r + 20q = -4a - 8$$

Given these revisions to the second equation let's look again at the entire system of equations.

$$-12r + bq = 4$$
$$-12r + 20q = -4a - 8$$

On the SAT® test I would recommend solving for a at this point. However, as a mathematician all these negatives are bothersome. We are going to use an old math trick and multiply both equations by negative 1. This will unravel the mess of negative signs while maintaining the equivalency of the equations. I'll show the equations after multiplying them by negative 1, you can work through the math on your own. Notice, that you all you need to do is change the sign of every term in each equation.

$$12r - bq = -4$$
$$12r - 20q = 4a + 8$$

Since this system of equations has an infinite number of solutions, each term in the upper equation must be equal to its respective term in the lower equation. Which results in the following two equations:

$$-bq = -20q$$
$$-4 = 4a + 8$$

This question is only asking us to solve for a. Use the lower equation from the two above to find a. Start by subtracting 8 from both sides.

$$-4 - 8 = 4a + 8 - 8$$
$$-12 = 4a$$

Now all that is left to do in order to get a by its lonesome is to divide the resulting equation by 4 in other words multiply the equation by $\frac{1}{4}$.

$$\frac{1}{4}(-12 = 4a)$$
$$-\frac{12}{4} = \frac{4}{4}a$$
$$-3 = a$$

The correct answer is -3.

Chapter 4

Passport to Advanced Math

4.1 Systems of Equations

Problem 1

$$-6x^2 + 3y^2 = 168$$
$$y = -3x$$

If (x, y) is a solution to the system of equations above, what is the value of x^2?

Solution and Explanation

Conveniently enough the bottom equation is already solved for the variable y. Since the question is asking for the solution in terms of x^2 we are going to treat that value as its own variable and will not try to simplify it to x. Plug the bottom equation into the top one, giving us one equation with only the variable x^2, after preforming some multiplication.

$$-6x^2 + 3(-3x)^2 = 168$$
$$-6x^2 + 3(9x^2) = 168$$
$$-6x^2 + 27x^2 = 168$$

Combine the like terms.

$$21x^2 = 168$$

Divide the entire equation by 21, in other words multiply the entire equation by $\frac{1}{21}$

$$\frac{1}{21} \times (21x^2 = 168)$$
$$\frac{21}{21}x^2 = \frac{168}{21}$$
$$x^2$$

Problem 2

$$2t + s = 2$$
$$(t-1)^2 - 4(t-1) - 15 = s$$

If (t, s) is a solution of the system of equations above and $s > 0$, what is the value of s?

Solution and Explanation

The SAT® test at this level is built to really test your mathematical maturity. The only real way to solve problems like this in a timely manner is to practice problems of this difficulty over and over again. At the same time as you are practicing them you need to use different techniques to solve the same equation. Back in engineering school the saying went, "there are many ways to skin a cat." Which means there a lot of ways to solve a problem. On the SAT® test it is going to come down to the best and most efficient way to skin the cat, i.e. to solve the problem. For this problem the best path to the solution is using a substitution method. Again, you will only be able to notice that quickly and under pressure when you have practiced it several times.

As we walk through this problem pay attention to the steps and method of thinking. After you have completed it, I recommend going over it at least 3 times or until you can do it without looking at the solution.

Looking at the two equations in the problem we see that s appears by itself in both equations. The bottom one is already solved for s. Which means it might be tempting to take that equation and immediately plug it into the upper equation. However, if you are patient in this situation and take a closer look at the upper equation and solve for s you will notice that it greatly simplifies your path to the solutions. Let's rearrange the top equation and solve for s by subtracting $2t$ from both sides.

$$2t - 2t + s = -2t + 2$$
$$s = -2t + 2$$

The right side of the equation almost, but not quite, looks like the $(t-1)$ term in the lower equation. If we were to factor out a negative 2 from the $-2t$ and 2 terms we could transform it into a $(t-1)$ term. That would look as follows:

$$s = -2(t-1)$$

On your own scratch paper, distribute the -2 back into the parenthesis to see that the two different expressions on the right side of the equation really are equivalent.

Now that we have solved the upper equation in terms of s we can plug it into the lower equation. That looks as follows:

$$(t-1)^2 - 4(t-1) - 15 = -2(t-1)$$

At this point we have a $(t-1)$ term on both the left and the right side of the equation. Whenever you are dealing with equations it always best to combine like terms; this case is no different. Let's bring the term from the right side of the equation to the left side by adding $2(t-1)$ to both sides of the equation.

$$(t-1)^2 - 4(t-1) - 15 + 2(t-1) = -2(t-1) + 2(t-1)$$
$$(t-1)^2 - 4(t-1) - 15 + 2(t-1) = 0$$

Now combine the two $(t-1)$ terms to get the following equation:

$$(t-1)^2 - 2(t-1) - 15 = 0$$

Here is where the substitution comes into play. Let's take a variable and call it u and set that variable equal to $(t-1)$. Watch what happens to the equation when we do that:

$$(t-1)^2 - 2(t-1) - 4 = 0 \qquad \text{with } u = (t-1)$$
$$u^2 - 2u - 15 = 0$$

Using the substitution method, we get a quadratic equation. This can be easily solved by using the quadratic formula. But I advise before jumping into the quadratic formula that you try to factor out the polynomial by finding two numbers that add or subtract to -2 and multiply to -15. Take about 15 seconds to think about that before looking at the next sentence for the solution. The two numbers we can use to factor this quadratic equation are -5 and 3. So that the quadratic equation in the factored form looks as follows:

$$(u-5)(u+3) = 0$$

Using the above method to find the factored form of a quadratic equation is often an easier method of solving the equation than running through the quadratic formula. However, if you find yourself taking more than 15 to 20 seconds to find the two numbers you need, I would recommend jumping into the quadratic formula, because it is a certain way to get the solutions for u.

Now that we have the quadratic equation in factored form, we can easily solve for u by setting each factor to 0.

$$u - 5 = 0$$
$$u + 3 = 0$$

The top equation can be solved for u by adding 5 to both sides.

$$u - 5 + 5 = 0 + 5$$
$$u = 5$$

The bottom equation can be solved for u by subtracting 3 from both sides of the equation.

$$u + 3 - 3 = 0 - 3$$
$$u = -3$$

Given the equations above we can see that both the solutions for u are $u = 5$ and $u = -3$. Now, we cannot forget that u is actually equal to $(t-1)$. So that the two equations in the sentence prior to this one become:

$$t - 1 = 5$$
$$t - 1 = -3$$

If we add 1 to both sides of both of the equations above. We can solve for t. I will write down the solutions, you should see if you can follow through with the math.

42

$$t = 6$$
$$t = -2$$

Now that we have values for t we can find the values for s. Using the first equation that we solved for s above will be the easiest way to solve for s given values of t. The equation was:

$$s = -2t + 2$$

For $t = 6$ the value of s is:

$$s = -2(6) + 2 = -12 + 2 = -10$$
$$s = -10$$

This is not the solution we are looking for because the question is asking for values of s greater than 0. Let's take a look at the value of s given $t = -2$.

$$s = -2(-2) + 2 = 4 + 2 = 6$$
$$s = 6$$

The correct answer is $s = 6$.

4.2 Operations with Polynomials and Rewriting Expressions

Problem 1

$$(q^2 + cq + 3)(q - 2) = q^3 + q^2 - 3q - 6$$

In the equation above, c is a constant. If the equation is true for all values of q, what is the value of c?

A) 4

B) 3

C) 6

D) 2

Solution and Explanation

When you see a polynomial combined with the phrase, "equation is true for all values of *the variable*" there is a very good chance that you are working with a coefficient comparison problem. Which means we need to set up the polynomials on both side of the equation to be of a similar form. The polynomial on the right side of the equation is already written in standard form and is of the third degree. Our first goal in this problem is to multiply out the left side of the equation. This is accomplished by distributing q^2 across $(q - 2)$ and then doing the same thing for cq and 3.

$$(q^2 + cq + 3)(q - 2) = q^3 - 2q^2 + cq^2 - 2cq + 3q - 6$$

Now that the polynomial is multiplied out, we still need to combine the like terms so that it matches the polynomial on the right side of the equation from the question. This can be easily accomplished by factoring q^2 and q out of their respective terms in the polynomial.

$$q^3 - 2q^2 + cq^2 - 2cq + 3q - 6 = q^3 + (c - 2)q^2 + (3 - 2c)q - 6$$

For clarity the multiplied-out polynomial from the left side of the equation in the problem looks as follows:

$$q^3 + (c - 2)q^2 + (3 - 2c)q - 6$$

Which is comparable to the right side of the equation from the problem.

$$q^3 + q^2 - 3q - 6$$

To find the solution we can either compare the q^2 coefficient or the q coefficient. Both ways will bring us to the same result. In this case it is easier when we compare the coefficient for the q^2 term, which are $(c - 2)$ for the left side of the equation, and 1 for the right side of the equation. Set those two coefficients equal to one another and solve for the variable c.

$$c - 2 = 1$$

Add 2 to both sides of the equation.

$$c - 2 + 2 = 1 + 2$$
$$c = 3$$

Because $c = 3$, the correct answer is B. Notice that if you compare the coefficient for the q term you will come to the same answer. Try it out on your own for verification.

Problem 2

Which expression is equivalent to $2x^3y - 4xy^2 + x^3y$?

A) $2x^3y - 4xy^2$

B) $-x^6y$

C) $-x^7y^4$

D) $3x^3 - 4xy^2$

Solution and Explanation

The solution to this problem can be found by adding the coefficients of like terms. In this problem the only like term is x^3y. It has the coefficients 2 and 1, which means the total coefficient will be $2 + 1 = 3$. Let's take a look at this addition in a more formal manner to come to the final answer.

$$2x^3y - 4xy^2 + x^3y = (2+1)x^3y - 4xy^2 = 3x^3y - 4xy^2$$

The correct answer is D.

Problem 3

$$\frac{12t+3}{t-2}$$

Which mathematical expression is equivalent to the one written above?

A) $12 - \dfrac{3}{2}$

B) $12 + \dfrac{3}{2}$

C) $12 + \dfrac{27}{t-2}$

D) $12t + \dfrac{3}{t-2}$

Solution and Explanation

This problem puts to use one of the least used, but most essential, tools in the mathematician's tool belt. That is, knowing how to add 0 to an expression. Normally when you are working with equations you need to abide by the rule that whatever you do to one side of the equation you must do to the other side. But when you want to manipulate individual expressions on their own, you are only allowed to multiply by 1 or add 0. In this case you need to add 0 to the expression in such a way that you can cancel out the sum in the denominator.

Pay attention closely and the process will become clear as you see how the solution unfolds. Now, if you add $\dfrac{24}{t-2}$ and then subtract $\dfrac{24}{t-2}$ you will get the following sum:

$$\frac{12t+3}{t-2} - \frac{24}{t-2} + \frac{24}{t-2}$$

Because the fractions share a common denominator you can combine the terms by adding all the numerators in a single fraction. The expression above will then look as follows:

$$\frac{12t+3-24+24}{t-2}$$

Notice that at any time you can cancel out the 24's so that the expression is exactly the same as the one in the problem. However, here you can use the 24 to your advantage. To clear things up move the -24 over to the right of the 12t, so that the expression looks as follows:

$$\frac{12t-24+3+24}{t-2}$$

Now comes the real trick of the whole problem. You can factor a 12 out of $12t-24$ and you will get $12(t-2)$. Notice the sum $(t-2)$ is the same sum as in the denominator. Look how the expression looks after factoring out the 12:

$$\frac{12(t-2)+3+24}{t-2}$$

The next step for simplification is to add the 3 and the 24 to 27. After that separate the fraction into two fractions.

$$\frac{12(t-2)+27}{t-2} \longrightarrow \frac{12(t-2)}{t-2} + \frac{27}{t-2}$$

At this point it is possible to cancel out the sum $t-2$ from the left side inequality. After performing that operation, you will come to the final solution.

$$12 + \frac{27}{t-2}$$

Learning and understanding this trick is the difference between not scoring 800 and scoring 800 on the SAT math section. If you develop your mathematical understanding to know how to implement the addition by 0 for simplification of fractions, you are well on your way to mathematical accomplishments far beyond a perfect SAT® score. I recommend you review this problem a couple times until you understand it fully.

The correct answer is C.

4.3 Quadratic Functions and Equations

Problem 1

Tim is riding his skateboard at v feet per second. All of a sudden a cow is standing in his way and he has no choice but to stop. In his shock it takes him 1.2 seconds to begin braking. It takes him another $\frac{v}{36}$ seconds to brake completely, at an average speed of $\frac{v}{3}$ feet per second. If it takes Tim 45 feet, from the time that he saw the cow, to come to a complete stop, which of the following equations can be used to find the value of v?

A) $v^2 + 108v - 2,430 = 0$

B) $v^2 + 108v - 4,860 = 0$

C) $v^2 + 129.6v - 2,430 = 0$

D) $v^2 + 129.6v - 4,860 = 0$

Solution and Explanation

This problem is a dive back into the basic physical relationship between speed, time and distance. The form that is most helpful for us in this equation is the following form of the relationship.

$$distance = speed \times time$$

This problem presents information in sets of distance. Pay close attention and try to relate my groupings to the problem.

First, we are given a *speed* of travel, v feet per second, and the *time* taken to react before braking, 1.2 seconds.

Second, we are given the *time* it took to brake, $\frac{v}{36}$ seconds, and the average *speed* during braking, $\frac{v}{3}$ feet per second.

Last, we are given the total *distance* it took Tim to stop, 45 feet.

Looking at this information let's see what we have to work with from every chunk of information. In the first two chunks we have time and speed, which means we also know the distance because we have the equation form the physical relationship, $distance = speed \times time$. In the final chunk of information, we are only given the distance. From the distance we are not able to calculate the speed or time because we have two unknowns and only one equation. This means that the only consistent piece of information we have from every all of the information is the distance.

So, we are going to write the equation in terms of distance. We will add the reaction distance traveled, before braking began, to the distance traveled while braking. The sum of those two distances will equal the total distance of 45 feet.

The reaction distance traveled is the speed, v feet per second, multiplied by the reaction time of 1.2 seconds. That looks as follows with the units left off.

$$1.2v$$

The distance traveled during braking is the average braking speed, $\frac{v}{3}$ feet per second, multiplied by the time it took to break, $\frac{v}{36}$ seconds. That looks as follows with the units left off.

48

$$\frac{v}{3}\frac{v}{36} = \frac{v^2}{108}$$

Now we can build a sum given those two distances and set it equal to 45 feet.

$$\frac{v^2}{108} + 1.2v = 45$$

This is called a quadratic equation because the variable v is raised to the second power. Quadratic equations are written in standard form as follows:

$$v^2 + bv + c = 0$$

When we look at the standard form of the quadratic equation and compare it to the equation we wrote in our problem, we see that two things need to be done.

- Get a 0 on the right side of the equation.

- Get a coefficient of 1 in front of the v^2 term

In order to get a zero on the right side of the equation, subtract 45 from both sides of the equation.

$$\frac{v^2}{108} + 1.2v - 45 = 45 - 45$$

$$\frac{v^2}{108} + 1.2v - 45 = 0$$

To get the coefficient of the v^2 term to 1 multiply the entire equation by 108.

$$108(\frac{v^2}{108} + 1.2v - 45 = 0)$$

$$\frac{108}{108}v^2 + 108 \times 1.2v - 108 \times 45 = 108 \times 0$$

$$v^2 + 129.6v - 4,860 = 0$$

The correct answer is D.

Problem 2

What is the solution to the equation $x^2 - 5 = 3x$?

A) $\dfrac{-3 \pm \sqrt{29}}{2}$

B) $\dfrac{-3 \pm \sqrt{11}}{2}$

C) $\dfrac{3 \pm \sqrt{29}}{2}$

D) $\dfrac{3 \pm \sqrt{11}}{2}$

Solution and Explanation

The type of equation in the above problem is a quadratic equation. That is because the variable, x, is to the second power. A quadratic equation can be solved with the quadratic formula. In general, the quadratic formula looks as follows:

$$x = \frac{-b \pm \sqrt{b^2 - 4ac}}{2a}$$

This formula is extremely important, so important that I would recommend taking a minute every day to write it down until you have committed it to memory.

The variables a, b and c in the equation refer to the coefficients found in the general form of the quadratic equation, which looks as follows:

$$ax^2 + bx + c = 0$$

The equation written in the problem is not in the general form. In order to solve this problem, it needs to be rearranged to the correct form. We can accomplish this by subtracting $3x$ from both sides.

$$x^2 - 3x - 5 = 3x - 3x$$
$$x^2 - 3x - 5 = 0$$

After rearranging the equation, we can see that from the general form $a = 1$, $b = -3$ and $c = -5$. Those values can be plugged into the quadratic formula which will result in the following:

$$x = \frac{-(-3) \pm \sqrt{(-3)^2 - 4(1)(-5)}}{2(1)}$$

This solution needs a little bit of work in order to simplify the terms. First, we'll perform the multiplication and then follow that up with addition and subtraction where possible.

$$x = \frac{3 \pm \sqrt{9 + 20}}{2}$$
$$x = \frac{3 \pm \sqrt{29}}{2}$$

The correct answer is C.

4.4 Solving Rational Equations

Problem 1

If the expression $\frac{9q^2}{3q-1}$ is written in the equivalent form $\frac{1}{3q-1}+Z$, what is Z in terms of q?

Solution and Explanation

When someone tells me that two things are equivalent or equal, the first thing I tend to do is write out their equality with a mathematical equation. Given the question above, write out the following mathematical equation:

$$\frac{9q^2}{3q-1} = \frac{1}{3q-1}+Z$$

To solve for Z in terms of q we need to get the single Z on one side all by itself, and all of the q terms on one side by themselves. To do this, move the $\frac{1}{3q-1}$ term to the left side without forgetting to change the sign.

$$\frac{9q^2}{3q-1} - \frac{1}{3q-1} = Z$$

Under most circumstances I would say to multiple the entire equation by the denominator on the left side of the equation, so that everything would be sitting in the numerator on both sides. In this case however, I will advise against that because it would lead to a quadratic equation in q. Instead combine both numerators under a common denominator and then pay close attention.

$$\frac{9q^2-1}{3q-1} = Z$$

In this case you will have to remember, the not so commonly used, mathematical rule called the difference of squares:

Difference of Squares

$$(a+b)(a-b) = a^2 + ab - ab - b^2 = a^2 - b^2$$

The inner terms of the difference of Squares always fall away. Here I'll emphasize that the only way you can see these "special" mathematical relationships is to practice them. Wake up in the morning and write out 3 difference of squares products while you are eating breakfast. Rearrange the products and place them into fractions so that you can become more comfortable with these mathematical rarities.

This equation has the relationship with the difference in squares, in that $a = 3q$ and $b = 1$, which in turn means that $a^2 - b^2 = (3q)^2 - 1^2 = 9q^2 - 1$. Now use the difference of squares relationship to rewrite the entire equation with the numerator in a factored form.

$$\frac{(3q+1)(3q-1)}{3q-1} = Z$$

The equation above is a SAT® miracle. The factor $3q-1$ appears in both the numerator and the denominator and therefore is cancelled out. Which leaves Z as linear equation in q.

$$3q+1 = Z$$

Problem 2

$$X(t) = \frac{t^4 + 3t^3 - 7t + 15}{(t-7)^2 + 6(t-7) + 9} \tag{4.1}$$

For what values of t is the function X undefined?

Solution and Explanation

A function is undefined when the denominator is 0. That is because you cannot, I repeat, *cannot* divide a number by 0.

Don't let the numerator of this function scare you. In order to solve this problem, you only need to set the denominator equal to 0 and solve for t. That looks mathematically as follows:

$$(t-7)^2 + 6(t-7) + 9 = 0$$

As the equation above is by itself, it would be quite a bit of work to solve for t. Notice however, that both times t appears, it is a part of the sum $t - 7$. When you notice something like this the best thing to do is to use a substitution variable. Call it u and set it equal to $t - 7$ so that the equation above becomes:

$$u^2 + 6u + 9 = 0$$

Before jumping immediately into the quadratic formula to solve this quadratic equation, see if you can factor the quadratic equation. Spend a maximum of 20 seconds to find numbers that add up to 6 and multiply to 9. Think about this before reading on. The number that meets the requirements is 3, this is the number you can use to factor the above quadratic equation. In its factored form the equation looks as follows:

$$(u+3)(u+3) = 0$$

Because $(u+3)$ is multiplied by itself, setting either $(u+3)$ equal to 0 will deliver the solution to this equation.

$$u + 3 = 0$$

In order to solve for u subtract 3 from both sides of the equation.

$$u + 3 - 3 = 0 - 3$$
$$u = -3$$

At this point it is very important to remember that u is only a substitution variable. Actually u is equal to $t - 7$; by plugging that into the place of u it frees you up to solve for the value of t.

$$t - 7 = -3$$

In order to solve the above equation for t add 7 to both sides of the equation.

$$t - 7 + 7 = -3 + 7$$
$$t = 4$$

If you plug $t = 4$ into the denominator of the function X from the problem, it will bring the denominator to 0. Therefore, making the function undefined for the value $t = 4$.

4.5 Function Notation

Problem 1

The function s is defined by $s(t) = kt^4 + 12t^3 - 44t + 30$, where k is a constant. In the ts-plane, the graph of s intersects the t-axis at the three points (-5,0), (1,0), and (q,0). What is the value of k?

Solution and Explanation

The function $s(t)$ above is a single variable function. You are probably more accustomed to seeing single variable functions written as $f(x)$. The truth is that there are a multitude of ways to annotate a single independent variable. If you for example see the function r written as $r(p)$, you need to recognize what the dependent and independent variables are.

- The independent variable of a function always lies inside the parenthesis.

- The dependent variable of a function is always directly to the left of the parenthesis.

Towards the end of the question you will see groupings of numbers within parenthesis. These are called coordinates. More importantly they are solutions to the function. It is important to not look at a function in terms of x and y, but rather in terms of the independent and dependent variables. Understanding this you can interpret the coordinates as follows:

- (independent variable, dependent variable)

    Example:

    (t,s)

While taking the actual SAT® test you are not only concerned with how to solve the problem, but also how to solve the problem most efficiently. This question is asking for the value of the unknown constant k. Recognize there is only *one* unknown and *one* equation! This means we only need to use one set of coordinates to find the solution. Looking at the function again, notice that there is a lot of multiplication going on, especially with the t^4 term. You should practice solving this problem with a different set of coordinates; however, for efficiency reasons select (1,0) to solve the equation k. To use this set of coordinates, replace every t in the equation with the number 1. Replace $s(t)$ with the value of 0. The equation will then look like:

$$0 = k \times 1^4 + 12 \times 1^3 - 44 \times 1 + 30$$

By performing the previous step, notice that you have just transformed an "advanced math" problem into an algebra problem. All that is left to do is to solve for k. Let's first rewrite the problem to get rid of all the powers so that we have a better view of what lies in front of us.

$$0 = k + 12 - 44 + 30$$

In order to solve for k move all the numerical terms to the left side of the equation, perform the arithmetic and see what remains. Remember that when a term moves from one side of the equation to the other, the sign needs to change.

$$-12 + 44 - 30 = k$$
$$44 - 42 = k$$
$$2 = k$$

The solution to this problem is $k = 2$. The important thing to take away from this problem is how to insert coordinates into a function along with the relationship between the function and its variables.

Problem 2

If $h(x) = \frac{1}{4}x + 3$ and $g(x) = h(x) + 1$, what is $g(8)$?

Solution and Explanation

The purpose of a function is to provide an output given an input. In this question we are given the input 8. We need to plug that into the function $g(x)$ to find the output. What the wonderful people at the SAT® committee have done here is embedded a function as a term within another function. We can crack this rather easily by replacing the $h(x)$ term in the $g(x)$ function with its algebraic expression. I'll write that out logically for you.

If $h(x) = \frac{1}{4}x + 3$ and $g(x) = h(x) + 1$, then $g(x) = \frac{1}{4}x + 3 + 1$.

Now let's take the input of 8 and plug it into the function to see what the output is.

$$g(8) = \frac{1}{4} \times 8 + 3 + 1 = \frac{8}{4} + 4 = 2 + 4 = 6$$

The answer to this problem is $g(8) = 6$.

Problem 3

$$s(t) = bt^4 - 13$$

For the function s defined above, b is a constant and $s(2) = 19$. What is the value of $s(-2)$?

A) 19

B) 0

C) -19

D) 2

Solution and Explanation

This problem is testing your ability to recognize that when numbers are raised to an even numbered power then it doesn't matter if the number being raised is positive or negative, the result will be the same.

If you know this rule, you can immediately recognize that the answer must be 19, because $2^4 = (-2)^4$ which in turn means that $s(2) = s(-2)$.

The correct answer is A.

4.6 Relationships Between Algebraic and Graphical Representations of Functions

Problem 1

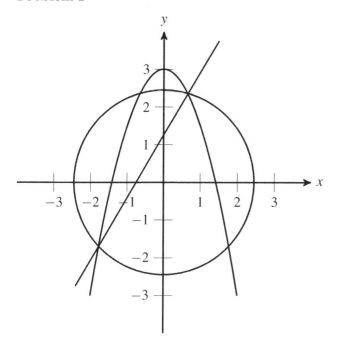

$$x^2 + y^2 = 6$$

$$y = -\frac{3}{2}x^2 + 3$$

$$y = \frac{5}{3}x + \frac{5}{4}$$

A system of three equations and their graphs in the *xy*-plane are shown above. How many solutions does the system have?

A) One

B) Two

C) Three

D) Four

Solution and Explanation

Using graphs to find the solutions to sets of equations really simplifies the path to the solution. Especially in this scenario where only the number of solutions is required. A solution occurs wherever there is an intersection of the lines graphed from the equations. For this problem all three lines intersect at two points. One time near the coordinates (0.7,2.4) and the second time near the coordinates (-1.8,-1.7).

The correct answer is B.

Problem 2

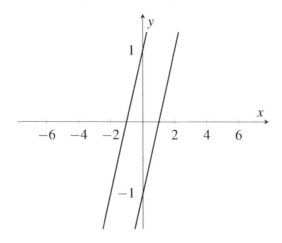

$y = x + 1$

$y = x - 1$

A system of two equations and their graphs in the xy-plane are shown above. How many solutions does the system have?

A) Zero

B) One

C) Two

D) Infinite

Solution and Explanation

This problem is presenting two linear equations with the same slope, but with different y-intercepts. Because the two lines both have the same slope, they are both moving in the same direction. Since the two lines are offset and have the same slope, they will never intercept. A solution of a system of equations only occurs where there is an intersection between all the curves from the system of equations at a single point. Because these lines will never intercept there can be no solution to this system of equations.

Before we advance from this problem, I would like to show how the scenario of no solutions to a system of equation looks like mathematically. Since both of the equations from the problem are already set equal to y, we can set the right side from each of the equations equal to one another to get one equation, which will allow us to solve for x. That equation will look as follows:

$$x - 1 = x + 1$$

In order to solve this equation, it is best to move all of the non-x terms to the right side of the equation and all of the x terms to the left side of the equation. Begin by adding 1 to both side of the entire equation in order to bring all the non-x terms to the right side.

$$x - 1 + 1 = x + 1 + 1$$
$$x = x + 2$$

When we try to collect all of the x terms onto the left side of the equation something unusual is going to occur. Namely, the x is going to disappear. Watch how that happens in the equation below, when we subtract x from both sides of the equation:

$$x - x = x - x + 2$$
$$0 = 2$$

The "equation" above is hardly that, because there is no equality between 0 and 2. When you see something like this when trying to solve systems of equation it is a definite signal that there are no solutions to the equations.

The correct answer is A.

Problem 3

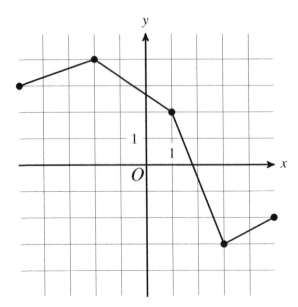

A complete graph of the function $g(x)$ is shown in the xy-plane. What is the x-value when the function $g(x)$ is at its maximum?

Solution and Explanation

This question is asking for the value of x when $g(x)$ is at its maximum. This can easily be confused with the y-value when the function $g(x)$ is at its maximum. Make sure to pay attention to this detail when solving the problem.

The maximum of the function is indicated by the word "Max" above the point on the curve located at (-2,4). See the graph below:

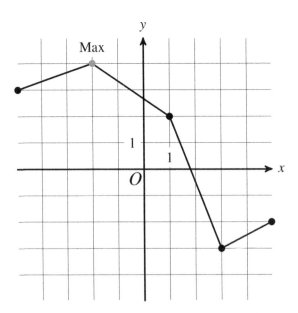

To find the x-value of this point draw a line vertically down from the point until it intersects with the x-axis. Once the intersection is made, count the number of vertical grid lines from the intersection point to the origin of the grid. If the intersection is on the left side of the origin, the number will be negative and if the intersection occurs on the right side of the origin, then the number for x will be positive.

For this problem after you draw a vertical line down from the maximum point of the function then you should count two vertical lines until you get to the origin. Because this intersection lies on the left side of the origin, 2 will be negative.

The *x*-value is -2 when the function $g(x)$ is at its maximum.

4.7 Exponential Functions, Equations, Expressions and Radicals

Problem 1

Which of the following is equivalent to $\left(\dfrac{3}{\sqrt[3]{x}}\right)^4$?

A) $3x^{\frac{4}{3}}$

B) $3x^{-\frac{4}{3}}$

C) $3x^{4+\frac{1}{3}}$

D) $3x^{4-\frac{1}{3}}$

Solution and Explanation

This problem boils down to whether or not you know the rules of exponents. The rules that you need for this equation are:

- Rule 1: $\sqrt[n]{a} = a^{\frac{1}{n}}$

- Rule 2: $\dfrac{1}{a^n} = a^{-n}$

- Rule 3: $(a^n)^m = a^{nm}$

The first thing we are going to do to the mathematical expression above is change the root into an exponent using Rule 1.

$$\left(\frac{3}{\sqrt[3]{x}}\right)^4 = \left(\frac{3}{x^{\frac{1}{3}}}\right)^4$$

Then we will move the x to the one third power, up into the numerator of the expression using Rule 2.

$$\left(\frac{3}{x^{\frac{1}{3}}}\right)^4 = \left(3x^{-\frac{1}{3}}\right)^4$$

The final step is to multiply the exponents $-\dfrac{1}{3}$ with 4 using Rule 3.

$$\left(3x^{-\frac{1}{3}}\right)^4 = 3x^{-\frac{4}{3}}$$

The correct answer is B.

Problem 2

$$x - 3 = \sqrt{3x + 45}$$

What is the solution set for the above equation?

A) $\{-3\}$

B) $\{12\}$

C) $\{-3, 12\}$

D) $\{3, -12\}$

Solution and Explanation

This problem requires you to be able to rearrange non-linear equations. For this problem we will need to remember the following rules of exponents:

- Rule 1: $\sqrt[n]{a} = a^{\frac{1}{n}}$

- Rule 2: $(a^n)^m = a^{nm}$

- Rule 3: $a^1 = a$

In order to solve this equation, we need to get rid of the square root on the right side of the equation. To do this we need to square both side of the equation.

$$x - 3 = \sqrt{3x + 45}$$

When we square both sides of the above equation we get:

$$(x - 3)^2 = (\sqrt{3x + 45})^2$$

To see how the square root disappears let's first change the presentation of the square root based on Rule 1.

$$(x - 3)^2 = \left((3x + 45)^{\frac{1}{2}}\right)^2$$

Next go ahead and use Rule 2 to multiply the two exponents on the right side of the equation above. Then simplify the expression as shown in Rule 3.

$$(x - 3)^2 = (3x + 45)^{\frac{2}{2}}$$
$$(x - 3)^2 = (3x + 45)^1$$
$$(x - 3)^2 = 3x + 45$$

With the right side of the equation cleared up, now we need to do something about that squared sum on the left side of the equation. Let's isolate the left side of the equation and perform the multiplication and distribution.

$$(x - 3)^2 = (x - 3)(x - 3) = x^2 - 3x - 3x + 9 = x^2 - 6x + 9$$

With the expanded form of the left side of the equation ready to go, we can now set it equal to the right side of the equation, and solve for x. First, let's look at the equation so that we can see what needs to be done.

$$x^2 - 6x + 9 = 3x + 45$$

Now we need to transform the equation above into the general form of the quadratic equation. Which means we combine all the like terms on the left side and get the right side to 0. We can accomplish this by subtracting 3x and 45 from both sides.

$$x^2 - 6x - 3x + 9 - 45 = 3x - 3x + 45 - 45$$
$$x^2 - 9x - 36 = 0$$

One way to solve this problem is to implement the quadratic formula. However, before going into that formula you can check for an easier solution by looking for two numbers which multiply together to be -36 and add or subtract to -9. Because the 36 is negative we know that one of the numbers will be negative and one will be positive. In this case the two numbers are -12 and 3. Now we can rearrange this quadratic equation into a factored form, given the two numbers just mentioned, in order to come to the answer. It looks as follows:

$$(x - 12)(x + 3) = 0$$

Being able to factor a polynomial like we just did is only a matter of practice. I would recommend cracking open the math book from your school and looking for the section on factoring polynomials. Knock out five to ten of these a day, and in no time, you will be able to spot these relationships with very little effort.

Back to the problem, in order for the equation above to be true, either one of the factors in the equation needs to be set to zero. By setting both of the factors equal to zero we will get two solutions to the quadratic equation.

$$x + 3 = 0$$
$$x - 12 = 0$$

Let's solve both of these equations simultaneously to receive two values for x. To solve the top equation, subtract 3 from both sides, to solve the bottom equation add 12 to both sides. Perform those operations on your own and I will present the solutions.

$$x = -3$$
$$x = 12$$

Now we have two potential solutions to the original equations. However, since we had to introduce a square in order to get rid of the square root there is the possibility that a solution has slipped in that might not actually solve the equation. This will become obvious when we plug in each solution into the original equation.

Starting with $x = -3$, plug that into the original equation and check if the solution is valid.

$$x - 3 = \sqrt{3x + 45}$$
$$-3 - 3 = \sqrt{3(-3) + 45}$$
$$-6 = \sqrt{-9 + 45}$$
$$-6 = \sqrt{36}$$
$$-6 = 6$$

The above equation is **not** true. That is because negative 6 does not equal positive 6. This means that $x = -3$ is not a valid solution to the original equation. Now let's try out the solution $x = 12$ to see if that results in a valid solution to the equation.

$$x - 3 = \sqrt{3x + 45}$$
$$12 - 3 = \sqrt{3(12) + 45}$$
$$9 = \sqrt{36 + 45}$$
$$9 = \sqrt{81}$$
$$9 = 9$$

The above equation is true. Which means that $x = 12$ is a valid solution to the equation. It is also the only solution to the original equation.

The correct answer is B.

Problem 3

For $j = \sqrt{-1}$, what is the sum $(3-4j)+(2+5j)$?

A) $5+j$

B) $5+9j$

C) $1+j$

D) $1+9j$

Solution and Explanation

The square root of negative one $\sqrt{-1}$ is known as the imaginary number and is commonly denoted with either an i or a j. In general, imaginary numbers are set up as follows:

- An imaginary number, $z = a+bj$, is made up of two components, they are called respectively the real and imaginary components.

    The real component, a, is a real number and *not* a product with the imaginary number.

    The imaginary component, bj, is a product made up of a real number and the imaginary number.

For two imaginary numbers $z = a+bj$ and $r = c+dj$. The sum of the imaginary numbers is the sum of the real components plus the sum of the imaginary components. In general, the addition looks as follows:

$$z+r = (a+bj)+(c+dj) = (a+b)+(cj+dj) = (a+b)+j(c+d)$$
$$z+r = (a+b)+j(c+d)$$

For this problem we can say that $z = 3-4j$ and $r = 2+5j$. Following the example above, we can add the two imaginary numbers from the problem as follows:

$$z+r = (3-4j)+(2+5j) = (3+2)+(-4j+5j) = 5+j$$
$$z+r = 5+j$$

The correct answer is A.

Problem 4

Which of the following complex numbers is equivalent to $\dfrac{4-8i}{3+4i}$? (Note: $i = \sqrt{-1}$)

A) $\dfrac{4}{3} + 2i$

B) $\dfrac{4}{3} - 2i$

C) $\dfrac{4}{5} + \dfrac{8}{5}i$

D) $\dfrac{4}{5} - \dfrac{8}{5}i$

Solution and Explanation

Division of complex numbers requires knowledge of the complex conjugate.

- For every complex number, $z = a + bi$, there is a complex conjugate, $z^* = a - bi$.

    The complex conjugate differs from the complex number by the sign in front of the imaginary term.

Other important rules of complex numbers to remember are:

- $i^1 = i$

- $i^2 = -1$

- $i^3 = (i^2)i = (-1)i = -i$

- $i^4 = (i^2)i^2 = (-1)(-1) = 1$

With all that being said now you have all the tools necessary to perform the division. We need to take the complex number and multiply both the numerator and the denominator by the complex conjugate of the denominator. That will looks as follows:

$$\left(\frac{4-8i}{3+4i} \right) \frac{3-4i}{3-4i}$$

At this point you can perform the normal fraction multiplication where you multiply the numerator of the left fraction by the numerator of the right fraction. The denominator of both the left and right fractions also need to be multiplied by one another. The problem with this multiplication is that there are sums in the numerators and denominators of both fractions. Which means you need to employ the F.O.I.L. method to complete the multiplication. Remember the F.O.I.L method?

- F. First: Multiply the first terms of each sum with each other.

- O. Outer: Multiply the outer terms of each sum with each other.

- I. Inner: Multiply the inner terms of each sum with each other.

- L. Last: Multiply the last terms of each sum with each other.

Start with multiplying the numerators using the F.O.I.L. method. Try doing the math on your own, before following along with the this:

$$(4-8i)(3-4i) = 12 - 16i - 24i + 32i^2$$

In the next step combine the imaginary terms and remember that $i^2 = -1$, this will result in:

$$12 - 16i - 24i + 32i^2 = 12 - 40i - 32$$

Now combine the real number terms to get the simplified form of the numerator.

$$12 - 40i - 32 = -20 - 40i$$

The above complex number is the new numerator after being multiplied by the complex conjugate of the denominator.

Now, let's look at the denominator. Again, use the F.O.I.L. method to multiply out the two sums. Try it out own your own before following along with the math.

$$(3+4i)(3-4i) = 9 - 12i + 12i - 16i^2$$

Notice that when multiplication between a complex number and its complex conjugate occurs, the imaginary terms cancel each other out. This will be always be the case with this kind of multiplication. In fact, if you ever multiply a complex number by its complex conjugate and the imaginary term remains, it is a very good indicator that something went wrong in your calculations.

Back to the problem, in the next math step cancel out the i terms and add together the real terms, keeping in mind that i^2 equals -1.

$$9 - 12i + 12i - 16i^2 = 9 + 16 = 25 \tag{4.2}$$

After completing the math for the numerator and the denominator. You can now build the fraction and simplify in terms of the possible solutions. The divided complex number looks as follows:

$$\frac{20 - 40i}{25}$$

This fraction can be simplified further by building a fraction out of each term in the numerator with the 25 from the denominator.

$$\frac{20 - 40i}{25} = \frac{20}{25} - \frac{40}{25}i$$

The final step is to factor out the values in the numerators and denominators so that the fractions can be brought to their simplest form. Each number in the above fractions is a multiple of 5, which means the factor of 5 should cancel out from each fraction.

$$\frac{20}{25} - \frac{40}{25}i = \frac{(4)5}{(5)5} - \frac{(8)5}{(5)5}i$$

All the fives cancel out, leaving

$$\frac{4}{5} - \frac{8}{5}i$$

The correct answer is D.

4.8   More Complex Problems in Context

Problem 1

If an object with the rotational inertia I is moving with the angular velocity ω, the object's rotational kinetic energy K is given by the equation $K = \frac{1}{2}I\omega^2$. If the moment of inertia of the object is halved and the angular velocity is doubled, how does the rotational kinetic energy change?

   A) The rotational kinetic energy is unchanged

   B) The rotational kinetic energy is quadrupled (multiplied by a factor of 4)

   C) The rotational kinetic energy is halved.

   D) The rotational kinetic energy is doubled.

Solution and Explanation

The question is asking how the rotational kinetic energy of an object changes when its moment of inertia, I, is multiplied by $\frac{1}{2}$ and the rotational velocity, ω, is multiplied by 2. In order to see how it changes we need to change the I in the equation to $\frac{1}{2}I$ and the ω to 2ω. Also, change the adjusted kinetic energy variable to E, so that we can differentiate it from the K in the original equation. The new equation looks as follows:

$$E = \frac{1}{2}\frac{1}{2}I(2\omega)^2$$

Right away there are a few things that we can simplify. Before the I there is a half multiplied by another half. That will become a quarter.

$$E = \frac{1}{4}I(2\omega)^2$$

The square above the parenthesis containing 2ω can be distributed to the elements of the product within the parenthesis.

$$E = \frac{1}{4}I4\omega^2$$

The $\frac{1}{4}$ cancels out the 4 in front of the ω. Performing that operation will bring E to its most simplified form.

$$E = I\omega^2$$

Given that $K = \frac{1}{2}I\omega^2$, a good mathematical eye will notice that the new kinetic energy E is twice as large as K. You could also say that K is half as much as E. On the SAT® test it would be best if you are at a mathematical level that you can notice this right away. That can be accomplished with practice, practice and more practice.
If you were unable to see the relationship between E and K, then there is a simple mathematical solution to find the relationship between E and K. We know that the two quantities will differ by an unknown multiple. Call the unknown multiple c and set up the following equation:

$$E = cK$$

Then plug in the mathematical expressions for each kinetic energy and solve the algebraic expression for c.

$$I\omega^2 = \frac{c}{2}I\omega^2$$

The product $I\omega^2$ is in the numerator on both sides of the equation. If we divide the entire equation by this product it will reduce to 1.

$$\frac{1}{I\omega^2}(I\omega^2 = \frac{c}{2}I\omega^2)$$
$$\frac{I\omega^2}{I\omega^2} = \frac{cI\omega^2}{2I\omega^2}$$
$$1 = \frac{c}{2}$$

to finish solving for c multiply the entire equation by 2.

$$2(1 = \frac{c}{2})$$
$$2 = \frac{2}{2}c$$
$$2 = c$$

Given this value for c the relationship between E and K becomes more visible. The equation E = cK becomes E = 2K. Here it is very clear that the new kinetic energy is twice as large as the one presented in the problem.

By halving the moment of inertia, I, and doubling the angular velocity, ω. The rotational angular velocity is doubled.

The correct answer is D.

Problem 2

$$\frac{\left(1+\dfrac{\pi}{64}\right)^{1.4}\left(16x-\dfrac{11}{5}y\right)^{4}}{1-\dfrac{\pi}{32}}T=\frac{1.6}{80\times10^{9}\left(\dfrac{\pi}{32}\right)(0.002)^{5}}P$$

An engineer is trying to figure out the relationship of P in terms of T,x,y as part of a project to build a vehicle. How would he rewrite the equation above in terms of P?

A) $P=\dfrac{80\times10^{9}\left(\dfrac{\pi}{32}\right)(0.002)^{5}\left(1+\dfrac{\pi}{64}\right)^{1.4}\left(16x-\dfrac{11}{5}y\right)^{4}}{1.6\left(1-\dfrac{\pi}{32}\right)}T$

B) $P=\dfrac{1.6\left(1+\dfrac{\pi}{64}\right)^{1.4}\left(16x-\dfrac{11}{5}y\right)^{4}}{80\times10^{9}\left(\dfrac{\pi}{32}\right)(0.002)^{5}\left(1-\dfrac{\pi}{32}\right)}T$

C) $P=\dfrac{\left(1+\dfrac{\pi}{64}\right)^{1.4}\left(16x-\dfrac{11}{5}y\right)^{4}}{1-\dfrac{\pi}{32}}T$

D) $P=\dfrac{\left(1+\dfrac{\pi}{64}\right)^{1.4}\left(16x-\dfrac{11}{5}y\right)^{4}}{1-\dfrac{\pi}{32}}T-\dfrac{1.6}{80\times10^{9}\left(\dfrac{\pi}{32}\right)(0.002)^{5}}$

Solution and Explanation

This problem may look intimidating, but it can be solved with relatively few steps. The important thing to remember when you are rearranging equations with fractions on either side of an equation is the following:

- Whatever is in the numerator one one side of the equation will be in the denominator on the other side of the equation.

- Whatever is in the denominator on one side of the equation will be in the numerator on the other side of the equation.

In this case we are trying to solve the equation in terms of P. Which means we need to move everything that is not P from the right side of the equation over to the left side. If you use the tips above, this problem can be solved in two simple steps. In that you take everything, except P, that is in the numerator on the right side of the equation and send it over to the denominator on the left side. Looking at the problem, you can see that the value in the numerator on the right side is 1.6. Bring this over to the denominator on the left side and the new equation looks as follows:

$$\frac{\left(1+\dfrac{\pi}{64}\right)^{1.4}\left(16x-\dfrac{11}{5}y\right)^{4}}{1.6\left(1-\dfrac{\pi}{32}\right)}T=\frac{P}{80\times10^{9}\left(\dfrac{\pi}{32}\right)(0.002)^{5}}$$

70

Now all that is left to do is to bring over the product, $80 \times 10^9 \left(\dfrac{\pi}{32} \right) (0.002)^5$, from the denominator on the right side of the equation over to the left side of the equation. That looks as follows:

$$\frac{80 \times 10^9 \left(\dfrac{\pi}{32} \right) (0.002)^5 \left(1 + \dfrac{\pi}{64} \right)^{1.4} \left(16x - \dfrac{11}{5}y \right)^4}{1.6 \left(1 - \dfrac{\pi}{32} \right)} T = P$$

The tips above become a little bit more complicated when there are sums of fractions on either side of an equation. When that is the case, I would recommend using a systematic algebraic approach like the methods used throughout most of the other problems in this book.

However, if you see *one* fraction on the left side and *one* fraction on the right side of an equation, then I would definitely recommend putting these simple tricks to use.

The correct answer is A.

Problem 3

A bacterial colony grows at a rate of 27% per day. If the initial number of bacteria was 1,775, which of the following functions N models the number of bacteria in the colony after d days?

A) $1,775(1.27)^d$

B) $1,775(0.27)^d$

C) $1,775(1.73)^d$

D) $1.27(1,775)^d$

Solution and Explanation

This problem is asking you to develop a function which models the growth of bacteria in a colony after d days. The initial number of bacteria is 1,775. After one day it will increase by 27%. Mathematically you will want to multiply 1,775 by 27% in such a way that 1,775 becomes larger. That can be accomplished with the following function for N evaluated at $d = 1$:

$$N(d = 1) = 1,775(1.27)$$

The above function can technically also be written as:

$$N(d = 1) = 1,775(1.27)^1$$

Which means technically after the first day, the new number of bacteria is the initial amount of bacteria multiplied by the 27% increase raised to the first power.

After another whole day you will need to multiply the above expression again by 1.27 to account for the second day's growth. That will look as follows:

$$N(d = 2) = 1,775(1.27)(1.27)$$

The above function can also be written as:

$$N(d = 2) = 1,775(1.27)^2$$

Which means after the second day, the new number of bacteria is the initial number of bacteria multiplied by the 27% increase raised to the second power.

Do you see the pattern that is developing here? After the first day you raise the percentage multiplier to the first power. After the second day you raise the percentage multiplier to the second power. Which implies that after the dth day you will raise the percentage multiplier to the dth power. That can be expressed in functional form as follows:

$$N(d) = N = 1,775(1.27)^d$$

The correct answer is A.

Chapter 5

Problem Solving and Data Analysis

5.1   Ratio, Proportion, Units, and Percentage

Problem 1

A large kitchen bowl was filled with 160 red and green marbles. The ratio of green to red marbles was 7 to 1. How many red marbles were in the bowl?

A) 16

B) 20

C) 144

D) 140

Solution and Explanation

The problem states that the ratio of green to red marbles was 7 to 1. Another way to read this ratio is that for every 7 green marbles there will be 1 red marbles. Which in turn means that for every 8 marbles only 1 of them will be red or, one eighth of all the marbles are red.

Now that we know the fraction of red marbles in the bowl. we can write a mathematical equation expressing that the number of red marbles is one eighth the amount of the total number of marbles. That will look as follows:

$$\# \text{ of red marbles} = \frac{1}{8} \times \text{total } \# \text{ of marbles}$$

We know that the total number of marbles is 160. With that information we can solve for the number of red marbles in the above equation.

$$\# \text{ of red marbles} = \frac{1}{8} \times 160 = \frac{160}{8} = 20$$
$$\# \text{ of red marbles} = 20$$

The correct answer is B.

Problem 2

An engineer's technical drawing of a steel plate 1 inch on the paper represents 48 inches on the steel plate. If the steel plate measures 3 inches by 1.5 inches on the paper, what is the actual area of the steel plate in square feet?

A) 72

B) 4.5

C) 216

D) 10,368

Solution and Explanation

The first thing to do for this problem is to build a mathematical ratio of the measurements on paper compared to the real-life measurements. That is 1 inch to 48 inches. The next step is to take the written ratio and turn it into a fraction. When you are building a ratio from words the thing to notice is the word "to" (it also commonplace to see the words "for every" or "per" in the place of "to") in the written ratio. The number that comes before "to" in the written ratio will be placed in the numerator of the fraction and the word that comes after "to" will by default be placed in the denominator of the fraction. In this case we have 1 inch *to* 48 inches, so the fraction or ratio will look as follows:

$$\frac{1 \text{ inch}}{48 \text{ inches}}$$

Notice that the ratio above represents the length of the steel plate on the paper, let's call that p, to the length of the steel plate in real life, which we will call l. This can also be expressed mathematically and set equal to the fraction above.

$$\frac{p}{l} = \frac{1 \text{ inch}}{48 \text{ inches}}$$

In order to find out the area of the steel plate in real life we need to first figure out the length and width of the plate given only the measurements of length and width on the paper. We can do that easily enough by rearranging the ratios in the above equation to solve for l and plugging in the respective values of p for the length and width of the steel plate on the paper.

Because l is in the denominator of the fraction on the left side of the equation, it can require some algebraic gymnastics to solve it. To save yourself some time, when you are presented with fractions on either side of the equal sign in an equation and the variable you wish to solve for is in the denominator of one of those fractions, it is usually best practice to immediately take the inverse on both sides of the equations. Remembering, what you do to one side of an equation you must do to the other.

To take the inverse of a fraction just swap the numerator with the denominator and vice versa. Performing the inverse operation on the fractions in the equation above results in the following:

$$\frac{l}{p} = \frac{48 \text{ inches}}{1 \text{ inch}}$$

Now all that is left to do to solve for l is to multiply the entire equation by p.

$$p\left(\frac{l}{p} = \frac{48 \text{ inches}}{1 \text{ inch}}\right)$$

$$\frac{p}{p}l = \frac{48 \text{ inches}}{1 \text{ inch}}p$$

$$l = \frac{48 \text{ inches}}{1 \text{ inch}}p$$

At this point we can solve for the real-life length and width of the steel plate. But before we do that remember that the units of the area are supposed to be in square feet. We should convert the units in the numerator of the fraction on the right side of the equation from inches to feet. That can be accomplished by multiplying the right side of the equation by the ratio of 1 foot for every 12 inches. Let's look at that multiplication along with the factorization of the ratio.

$$l = \frac{48 \text{ inches}}{1 \text{ inch}}\left(\frac{1 \text{ foot}}{12 \text{ inches}}\right)p$$

$$l = \frac{12 \times 4 \text{ inches}}{1 \text{ inch}}\left(\frac{1 \text{ foot}}{12 \text{ inches}}\right)p$$

Take a look at the bottom equation from the two above. Notice that both the 12 and the unit of inches cancel out of the ratio. So that the resulting equation becomes:

$$l = \frac{4 \text{ feet}}{1 \text{ inch}}p$$

With the equation above we can solve for the real-life length and width of the steel plate. Let's start with the length of the plate, measured on the paper it is 3 inches, which means set $p = 3 inches$.

$$\text{actual plate length} = \frac{4 \text{ feet}}{1 \text{ inch}} \times 3 inches$$

Notice how the unit of inches drops out, leaving the answer in the unit of feet.

$$\text{actual plate length} = \frac{4 \text{ feet}}{1} \times 3$$

$$\text{actual plate length} = \frac{4 \times 3 \text{ feet}}{1}$$

$$\text{actual plate length} = 12 \text{ feet}$$

Follow the same process to get the width of the plate in feet, given that the paper measurement is 1.5 inches.

$$\text{actual plate width} = \frac{4 \text{ feet}}{1 \text{ inch}} \times 1.5 inches$$

$$\text{actual plate width} = \frac{4 \text{ feet}}{1} \times 1.5$$

$$\text{actual plate width} = \frac{4 \times 1.5 \text{ feet}}{1}$$

$$\text{actual plate width} = 6 \text{ feet}$$

With the actual plate length and width in feet we can finally calculate the actual area of the steel plate in square feet. With the equation:

$$\text{Area} = \text{Length} \times \text{Width}$$

Plugging in the values Length = 12 feet and Width = 6 feet, the area is:

$$\text{Area} = 12\text{feet} \times 6\text{feet} = 12 \times 6 \times feet \times feet = 72\text{square feet}$$
$$\text{Area} = 72\text{feet}$$

The correct answer is A.

Problem 3

The orbit of the plant Jupiter around the Sun is about 4.89 billion kilometers. It takes around 11.86 earth years for Jupiter to complete its orbit around the Sun. What is the approximate average speed of Jupiter's orbit around the Sun in meters per hour?

A) 1.13 billion

B) 47.2 million

C) 47.2 thousand

D) 4.72 million

Solution and Explanation

This question is asking for us to find out the speed of Jupiter's orbit around the Sun. The only numerical information we are given for this problem is the distance traveled and the time taken for Jupiter to make one complete orbit. Remember the equation for speed from physics is:

$$\text{speed} = \frac{\text{distance}}{\text{time}}$$

Now plug in the values from the question to build a first level equation for the speed of Jupiter. We know from the question that the distance traveled is 4.89 billion kilometers and the time taken to travel that distance is 11.86 years.

$$\text{speed} = \frac{\text{distance}}{\text{time}} = \frac{4,890,000,000 \text{ kilometers}}{11.86 \text{ years}}$$

This equation is a good start to solving the problem, but the units still do not line up with the units the problem is asking for. This speed is in kilometers per year, whereas the question is asking for a speed in the units of meters per hour.

In order to solve this problem, you need to remember that there are 1,000 meters in 1 kilometer, 52 weeks in a 1 year, 7 days in a week and 24 hours in a day. When building the unit conversion ratios make sure that the right units are in the numerator and denominator so that you can cancel out the units of kilometers and years.

Write the unit conversion ratio from kilometers to meters as follows:

$$\frac{1,000 \text{ meters}}{1 \text{ kilometer}}$$

Write the unit conversion for years to weeks as follows:

$$\frac{1 \text{ year}}{52 \text{ weeks}}$$

Write the unit conversion for days to weeks as follows:

$$\frac{1 \text{ week}}{7 \text{ days}}$$

Write the unit conversion for days to hours as follows:

$$\frac{1 \text{ day}}{24 \text{ hours}}$$

Now rewrite the speed equation above, but this time multiply the right side of the equation by the four unit conversion ratios. Notice how we set up the unit conversion ratios in such a way that the years and kilometers cancel out, so that the new units are meters per hour.

$$\text{speed} = \frac{4,890,000,000 \text{ kilometers}}{11.86 \text{ years}} \left(\frac{1,000 \text{ meters}}{1 \text{ kilometer}}\right) \left(\frac{1 \text{ year}}{52 \text{ weeks}}\right) \left(\frac{1 \text{ week}}{7 \text{ days}}\right) \left(\frac{1 \text{ day}}{24 \text{ hours}}\right)$$

$$\text{speed} = 47.2 \, \frac{\text{million meters}}{\text{hour}}$$

Problem 4

A clothing company buys its clothing from a wholesaler. The company usually sell its dresses at cost from the wholesaler plus 60%. During a sale, the company charged the whole sale cost plus 10%. If the sale price of the dress was $78, what was the usual price for the dress?

A) 137.28

B) 70.9

C) 113.44

D) 44.32

Solution and Explanation

For this problem we need to first determine the wholesale cost, given the sales cost, so that we can in turn figure out the usual cost. The sale price was 10% greater than the wholesale cost. This means mathematically that the sale price of the dress was a factor of 1.1 times more than the wholesale cost.

$$\text{sales cost} = 1.1(\text{wholesale cost})$$

In order to figure out the wholesale cost, the equation above needs to be rearranged so that the wholesale cost is all by itself. You can accomplish that by dividing the entire equation by 1.1, in other words multiplying the entire equation by $\frac{1}{1.1}$.

$$\frac{1}{1.1}(\text{sales cost} = 1.1\text{wholesale cost})$$
$$\frac{\text{sales cost}}{1.1} = \text{wholesale cost}$$

Since we know that the sales cost was $78 we can plug that into the equation above and solve for the wholesale cost.

$$\text{wholesale cost} = \frac{\text{sales cost}}{1.1} = \frac{78}{1.1} = 70.9$$
$$\text{wholesale cost} = 70.9$$

Using the same method to figure out what the sales cost was we can figure out what the usual cost was. We know that the usual cost was 60% more than the wholesale. That can also be written as, the usual cost was a factor of 1.6 times more than the wholesale cost.

$$\text{usual cost} = 1.6\text{wholesale cost}$$

Given that we already solved for the wholesale cost, at $70.9, we can now calculate the usual cost.

$$\text{usual cost} = 1.6\text{wholesale cost} = 1.6 \times 70.9 = 113.44$$
$$\text{usual cost} = 113.44$$

5.2 Interperting Relationships Presented in Scatterplots, Graphs, Tables, and Equations

Problem 1

Time (years)	Population
0	20×10^3
1	60×10^3
2	18×10^4
3	54×10^4

The table above shows the initial population (at time $t = 0$) of an emerging industrial city in the Midwest, along with the population growth over a 3-year period. Which of the following functions model the population, P(t), after t years?

A) $P(t) = 3,000t$

B) $P(t) = 20,000 + 3,000t$

C) $P(t) = 20,000(3^{-t})$

D) $P(t) = 20,000(3^t)$

Solution and Explanation

This problem is asking us to model the population growth, $P(t)$, from the information given in the table. Looking at the answer set you will notice that there are two different types of functions to model the tabulated data. The first two options show linear equations. If the data in the above problem represented linear growth, then the population in the right column would continually grow by equal amounts as the years in the column grow by equal amounts. However, since that is not the case here you can already exclude options A and B as possible answers.

Option C is multiplying 20,000 by 3^{-t}. Remembering the exponent rules, we can rewrite the function in option C to be:

$$P(t) = 20,000 \left(\frac{1}{3^t} \right)$$

We can check to see if this is correct by plugging in $t = 1$ into the function.

$$P(t = 1) = 20,000 \left(\frac{1}{3^1} \right)$$

There is really no need to do any more calculating here. This function valued at $t = 1$ is saying that after one year the population will be one third of the starting population. Looking at the data however, we see that the population does not decrease by one third but rather grows by a factor of 3. In fact, if we divide the population in year 2 by the population after year 1, we can see that in the second year the population grows again by a factor of 3.

$$\frac{P(t = 2)}{P(t = 1)} = \frac{180,000}{60,000} = 3$$

You can do the math for the population growth between years 2 and 3 and there you will also see that the population again grows by a factor of 3. Which means the answer must be D. As t increases every year the population will grow by another factor of 3. Let's take a look at the population growth for $P(t = 3)$ to see mathematically how the values will increase by a factor of 3 each year.

$$P(t = 3) = 20,000(3^3) = 20,000(3)(3)(3)$$

The correct answer is D.

Problem 2

A bank just opened behind the local convenience store; as part of an opening week deal, the bank is offering $2,500 certificates of deposit at a simple interest of 17% per year. The bank is selling certificates with terms of 1, 2, 3, or 4 years. Which of the following functions gives the total amount, m, in dollars, that a customer will receive when a certificate with a term of m years is finally paid?

A) $D = 2,500(1.17m)$

B) $D = 2,500(1 + 0.17m)$

C) $D = 2,500(1.17)^m$

D) $D = 2,500(1 + 0.17^m)$

Solution and Explanation

Two very important equations to remember, *especially* for the SAT test, are the equations for simple and compound interest. The equation for compound interest will be discussed in the next problem. The equation for simple interest is:

$$A = P(1 + rt)$$

Where:

- A is the total amount after a period of time

- P is the initial investment

- r is the interest rate expressed as a decimal

- t is the time period

If you wish to solve this equation quickly on the SAT® test, I would recommend writing it down along with the variable definitions, once every day, starting a couple of weeks before the test. If you do this, you can memorize the equation and then the problem becomes nothing more than plugging the values given from the problem into the equation.

Let's go ahead and define the numerical value for each variable in the simple interest equation. The total amount, A, is the variable D in this equation. We don't need to know the value for this variable because we can calculate it given that we have all the other variables available. The initial investment, P, is $2,500; the interest rate, r, is 17%, which in decimal form is 0.17 and the time period, t, is represented by an m in the problem statement and can be either 1, 2, 3, or 4 years, we will leave it in variable form knowing that the customer can choose how long they would like to leave their certificate with the bank. Given those numerical values for the variables we can rewrite the simple interest equations in terms of the variables from the problem as:

$$D = 2,500(1 + 0.17m)$$

The correct answer is B.

Problem 3

A bank just opened behind the local convenience store, as part of an opening week deal, the bank is offering $2,500 certificates of deposit at an interest of 17% per year, compounded semiannually. The bank is selling certificates with terms of 1, 2, 3, or 4 years. Which of the following functions give the total amount, m, in dollars, that a customer will receive when a certificate with a term of m years is finally paid?

A) $D = 2,500(1.17m)$

B) $D = 2,500(1 + 0.34m)$

C) $D = 2,500(1.17)^m$

D) $D = 2,500(1.085)^{2m}$

Solution and Explanation

Two very important equations to remember, *especially* for the SAT® test, are the equations for simple and compound interest. The equation for simple interest was discussed in the previous problem. The equation for compound interest is:

$$A = P\left(1 + \frac{r}{n}\right)^{nt}$$

Where:

- A is the total amount after a period of time

- P is the initial investment

- r is the interest rate

- t is the time period

- n is the number of times the interest is compounded per time period

If you wish to solve this equation quickly on the SAT® test, I would recommend writing it down along with the variable definitions, once every day, starting a couple of weeks before the test. If you do this, you can memorize the equation and then the problem becomes nothing more than plugging the values, given from the problem, into the equation.

Let's go ahead and define the values for the variables in the equation given the information from the problem. Let's go ahead and define the numerical value for each variable in the simple interest equation. The total amount, A, is the variable D in this equation. We don't need to know the value for this variable because we can calculate it given that we have all the other variables available. The initial investment, P, is $2,500; the interest rate, r, is 17% which in decimal form is 0.17; the number of times the interest is compounded per time period, n is 2 and the time period, t, is represented by an m in the problem statement and can be either 1, 2, 3, or 4 years, we will leave it in variable form knowing that the customer can choose how long they would like to leave their certificate with the bank.

Plugging these values into the compound interest equation, we get the following:

$$D = 2,500\left(1 + \frac{0.17}{2}\right)^{2m}$$

Although the above correction is correct, it still needs a little bit of arithmetic in order for it to exactly resemble a solution from above. Before doing the extra work, you can look at the list of solutions to the problem and see immediately that neither A nor B can be correct. The equations for those two options are linear equations, compared to the equation above which is an exponential equation. Looking closely at options C and D, notice that C can also not be the correct answer, because the exponent in that equation is m whereas the exponent in this equation is $2m$. Also, the base of the exponential expression in C is 1.17 which is $1 + 0.17$. The base of the exponential expression in the equation written above is $1 + \dfrac{0.17}{2}$ which cannot be equal to 1.17.

Let's work out the arithmetic inside the parenthesis of the compound interest equation, with the numerical values written above, to show that in fact D is the correct answer.

$$D = 2,500\left(1 + \frac{0.17}{2}\right)^{2m} = 2,500(1 + 0.085)^{2m} = 2,500(1.085)^{2m}$$
$$D = 2,500(1.085)^{2m}$$

The correct answer is D.

Problem 4

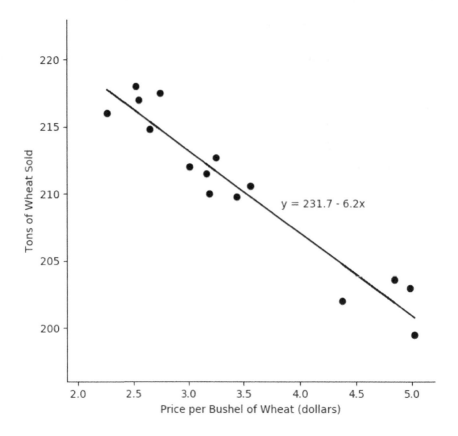

Tons of Wheat Sold

$y = 231.7 - 6.2x$

Price per Bushel of Wheat (dollars)

A local company buys wheat from the farmers in the area each week. The company also sets the price per bushel of wheat based on the amount of wheat the farmers have. The scatterplot above shows the price per bushel and the number of tons of wheat sold for 15 weeks, along with the line of best fit and the equation for the line of best fit.

For how many of the 15 weeks shown was the number of tons of wheat sold less than the amount predicted by the line of best fit?

A) 7

B) 6

C) 8

D) 9

Solution and Explanation

This question is asking for the number of weeks where the tons of wheat sold were less than the amount predicted. The values for tons of wheat sold are found on the vertical axis, which means when we are comparing the amount of wheat sold to the best fit line we will be moving in the vertical direction. To solve this problem, move along the best fit line from the top left down to the bottom right. While moving across the line continually look below the line to see if a point lies there. Every time you see a point count it, and keep a running tally of how many points you found there.

In the figure below is an example of counting off the points which lie below the best fit line.

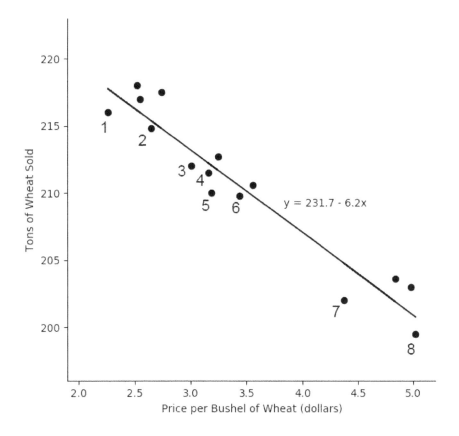

The correct answer is C.

Problem 5

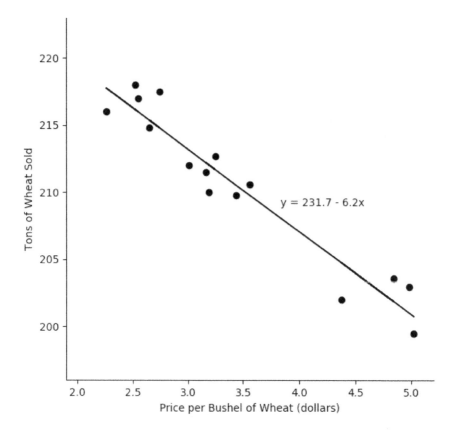

A local company buys wheat from the farmers in the area each week. The company also sets the price per bushel of wheat based on the amount of wheat the farmers have. The scatterplot above shows the price per bushel and the number of tons of wheat sold for 15 weeks, along with the line of best fit and the equation for the line of best fit.

According to the line of best fit, how many tons of wheat would the farmer need to harvest in one week to get a price per bushel of $4.00?

Solution and Explanation

The problem gives us a price per bushel and asks for the tons of wheat required to get that price according to the best fit line. The price per bushel is a value found on the horizontal or "x-axis", which means we can take the value of 4 and plug it into the x in the equation for the best fit line.

$$y = 231.7 - 6.2x = 231.7 - 6.2(4) = 231.7 - 24.8 = 206.9$$
$$y = 206.9$$

The variable y in this equation represents the values found on the vertical axis, or the tons of wheat sold. What this solution means is that in order for the farmer to get a price of $4.00 per bushel of wheat he would need to sell 206.9 tons of wheat.

Problem 6

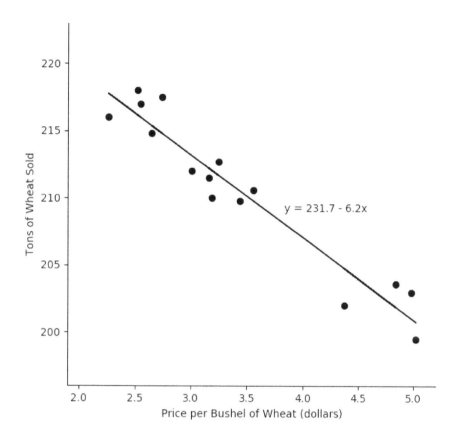

A local company buys wheat from the farmers in the area each week. The company also sets the price per bushel of wheat based on the amount of wheat the farmers have. The scatterplot above shows the price per bushel and the number of tons of wheat sold for 15 weeks, along with the line of best fit and the equation for the line of best fit.

What is the best interpretation of the meaning of the y-intercept of the line of best fit?

A) The price per bushel of wheat when 0 tons of wheat are sold.

B) The tons of wheat that need to be sold to get a price of $0.00 per bushel.

C) The maximum price per bushel a farmer can receive.

D) The tons of wheat a farmer would need to sell to receive a price per bushel of under $2.00

Solution and Explanation

The y-intercept in a linear equation can be found by setting the x value equal to 0. Setting the x value equal to 0 in the context of this problem would mean that the price per bushel of wheat would be 0 dollars.

This would be an unfortunate situation for the farmer because he would have 231.7 tons of wheat but wouldn't be able to get any money for his crop. This small exercise of interpreting the y-intercept shows that a best fit line may only be a valid interpretation of data over a range of x or y and not always for the entire range. When working with data it is always best to do a sanity check to see if the data makes sense within the range you are working in.

The correct answer is B.

Problem 7

A french fry factory has a new pneumatic air gun designed to shoot fries that are too small off of the production line. The production line manager estimates that 14,000 french fries pass the air gun every minute, and 176 of those fries are too small. The manager estimates the results of the new pneumatic air gun will be as shown in the table below.

	Pneumatic gun fires	Pneumatic gun does not fire	Total
French fry is too small	142	34	176
French fry is not too small	49	13,775	13,284
Total	191	13,809	14,000

According to the manager's estimates, if the air gun fires at a fry, what is the probability that the fry is *not* too small?

A) 0.35%

B) 0.24%

C) 1.03%

D) 25.7%

Solution and Explanation

The question is asking, of the instances that the pneumatic air gun fires, what is the probability that the fry is *not* too small. In one minute, the air gun will fire a total of 191 times according to the tabulated data. Of those 191 air shots at the fries 49 of them will be towards fries that are not too small. In other words, there is a 49 out of 191 chance that the pneumatic gun fires at a fry that is not too small. The word chance in the sentence prior is interchangeable with probability. Using this logic, we can build the following equation:

$$\text{Probability} = \frac{49}{191} = 0.257$$

Change this decimal into a percentage by multiplying the decimal by 100%.

$$\text{Probability} = 0.257 \times 100\% = 25.7\%$$

The correct answer is D.

Problem 8

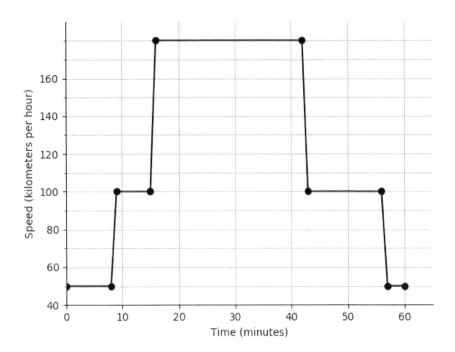

Everyday Lars drives for 60 minutes to get to work. On his way to the office he has to drive through cities, on country roads and on the autobahn. The graph above shows Lars' speed during the 60-minute trip. Which segment of the graph represents the times when Lars drives the slowest?

    A) The segment from (0,50) to (8,50) and (57,50) to (60,50)

    B) The segment from (9,100) to (15,100) and (43,100) to (56,100)

    C) The segment from (16,180) to (42,180)

    D) The segment from (8,50) to (9,100) and (56,100) to (57,50)

Solution and Explanation

The graph shown in the problem is a speed versus time graph. The nearly vertical lines represent times when Lars is accelerating. The horizontal lines represent times when Lars is traveling at a constant speed. Because the question is asking for when Lars drives the slowest, you are looking for horizontal lines on the graph where the speed is lowest. Notice that it cannot be the nearly vertical sections of the graph because those are section where Lars' speed is increasing or decreasing.

In the figure below you can see the sections of the lowest speed denoted by the dashed lines instead of solid lines on the graph.

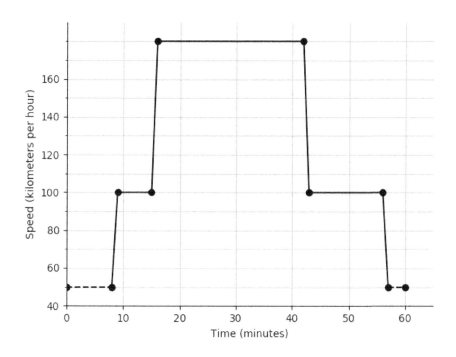

The dashed lines cover the range on the graph from (0,50) to (8,50) and (57,50) to (60,50).

The correct answer is A.

Problem 9

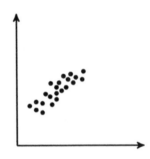

What is the statistical correlation shown in the scatter plot above?

A) Positive Correlation

B) Negative Correlation

C) Perfect Correlation

D) No Correlation

Solution and Explanation

The above problem is testing your skills on statistical correlations. Using linear correlations the goal is to draw a line through the points which best describes the tendency of the data points. There are several different types of statistical correlations and they can be described as follows:

- Positive Correlation: A group of data points which in general follow a positive mathematical slope from the bottom left to the upper right of the graph.

    This plot in this problem is an example of a positive correlation.

- Negative Correlation: A group of data points which in general follow a negative mathematical slope from the top left to the bottom right.

    See an example of negative correlation below

- No Correlation: A group of data points which seemingly follow no mathematical pattern what so ever.

    See an example of no correlation below

A perfect, strong or weak correlation is an indicator as to how tightly grouped the data points are.

- Perfect Correlation: A set of data points where every point lies exactly on a straight line.

    See an example of a perfect correlation below

    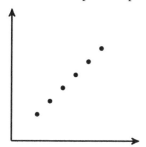

- Strong Correlation: A set of data points, which are tightly grouped around a straight line.

    See an example of a strong correlation below

    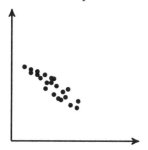

- Weak Correlation: A group of data points, which are loosely grouped around a straight line.

    See an example of a weak correlation below

    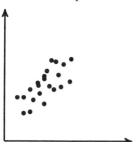

The extent to which strong and weak correlations differ from the perfect correlation may vary dramatically.

The correct answer is A.

5.3 Understand and Analyze Data Presented in a Table, Bar Graph, Histogram, Line Graph or Other Display

Problem 1

A young student decided to invest in a company in a European country. Every quarter he receives dividends from his investment. In order to pay the young student, the company must first convert the money to dollars, then pay a 2% charge on the converted dollar amount.

For the second quarter the company owed the student a dividend of 200.00 Euros. When they paid out, the bill from their bank charged 235.00 American Dollars.

What foreign exchange rate, in Euros per one American dollar, did the bank use for the company's charge? Round your answer to the hundredth decimal place.

Solution and Explanation

We need to perform the following steps to solve this problem:

- Figure out the uncharged amount owed to the student.

- Use the uncharged amount to solve for the conversion rate.

Since the conversion rate does not account for the 2% conversion charge paid by the company, we need to first solve for the uncharged amount. That is 2% written as 1.02 multiplied by some unknown amount of American dollars, d, equals the amount in American dollars charged to the company.

$$1.02d = 235.00$$

To solve for d divide the entire equation by 1.02, in other words multiply the entire equation by $\frac{1}{1.02}$.

$$\frac{1}{1.02}(1.02d = 235.00)$$
$$\frac{1.02}{1.02}d = \frac{235.00}{1.02}$$
$$d = 230.39$$

In order to find out the foreign exchange rate, r, of Euros to one American dollar, divide the amount owed to the student in Euros by the converted amount in American dollars, d.

$$r = \frac{200}{230.39} = 0.87$$

The correct answer is that the foreign exchange rate is 0.87.

Problem 2

The data below is used for Problems 2-4

Below is the relationship between the average height of a blade of grass and the amount of water the grass received, for nine different lawns. The linear best fit line is also shown.

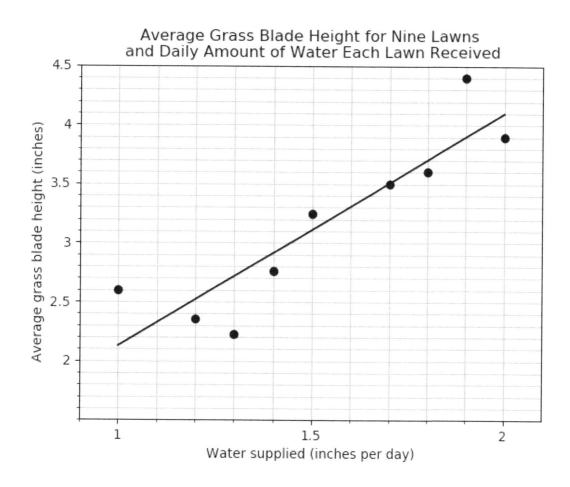

For how many lawns does the average grass blade height differ by more than 0.3 inches than the height predicted by the best fit line?

A) 6

B) 3

C) 4

D) 5

Solution and Explanation

The question is asking which average grass blades height differ by **more** than 0.3 inches. The average height of the blades of grass are denoted on the vertical axis. Each tick mark represents 0.1 inches. The best method for solving this problem is as follows:

- Run the end of your pencil along the best fit line.

96

- Every time you see a data point above or below the line, count the number of tick lines between the best fit line and the point.

- If three tick lines are between the best fit line and the data point, then you have a data point which meets the requirements of the question.

- Keep a running tally of all the points you find which meet the question requirements.

For this problem using the method described above there are three points. The first one is 1 inch of water per day on the horizontal axis. The second one is 1.3 inches of water on the horizontal axis. The third one is at 1.9 inches of water on the horizontal axis.

Since there are 3 points which meet the question requirements, the correct answer is B.

Problem 3

The data below is used for Problems 2-4

Below is the relationship between the average height of a blade of grass and the amount of water the grass received, for nine different lawns. The linear best fit line is also shown.

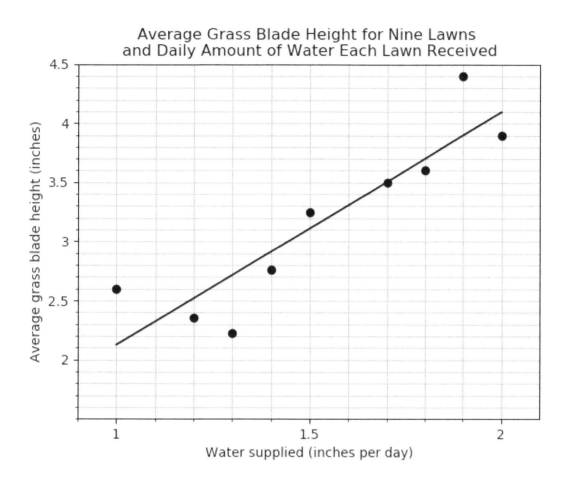

Which of the following is the best interpretation of the slope of the best fit line in the context of this problem?

A) The predicted height of a blade of grass in inches where a lawn received 0 inches of water per day.

B) The predicted amount of water in inches per day to get an average grass blade height of 0 inches.

C) The predicted amount of water in inches per day increase for every inch increase in grass blade height.

D) The predicted grass blade height increase for every inch per day of increase in water supplied.

Solution and Explanation

To solve this problem, you need to know that the slope is the rise over the run. In other words, the change in vertical position of a line over the change in horizontal position of a line.

The rise, the change in vertical position, in the context of this problem is *the increase in average grass blade height in inches.*

The change in horizontal position is often expressed in unit length, put simply the horizontal change will be measured as one unit of whatever unit is being used on the horizontal axis. In this case the unit is inches per day, so the run will be interpreted as *inch of water per day increase.*

When we put the rise over the run using the descriptions above we get the following: The slope, rise over run, in the context of this problem can be interpreted as *the increase in average grass blade height in inches* for every *inch of water per day increase.* comparing this definition to the options in the problem we see that it is very similar to the option D.

Taking a closer look at option C, notice that this option is describing the run over the rise. Which is the opposite to what the question is asking for. Option C is for this reason not the right answer.

Options A and B describe specific locations on the graph where the lawns receive 0 inches of water per day and the grass has a height of 0 inches. Both of these options refer to where the best fit line might intercept with the horizontal and vertical axes. Both of these options are also incorrect.

The correct answer is D.

Problem 4

The data below is used for Problems 2-4

Below is the relationship between the average height of a blade of grass and the amount of water the grass received, for nine different lawns. The linear best fit line is also shown.

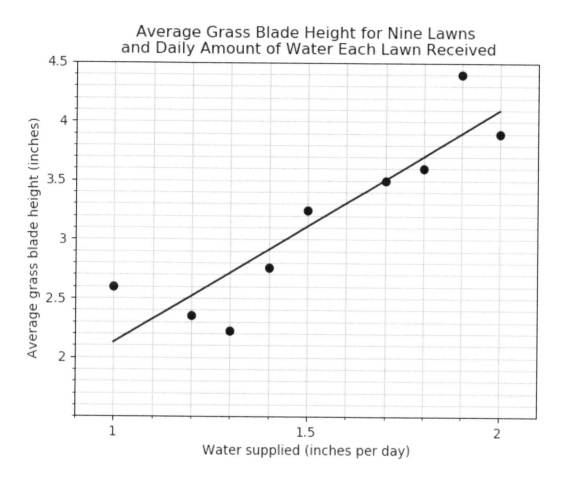

Based on the best fit line what is the average grass blade height for a lawn that receives 1.6 inches of water per day?

A) 3.3 inches

B) 3.4 inches

C) 3.2 inches

D) 3.5 inches

Solution and Explanation

This question requires you to find a point on the line given a location on the horizontal axis. Then take that point on the line and find its location on the vertical axis. For this problem take the end of your pencil to the tick line located at 1.6. Since each tick represents a spacing of 0.1 inches this position is one tick mark to the right of 1.5. Starting at the 1.6 tick mark perform the following steps:

- Move the pencil up to the best fit line.

- Notice which tick line crosses the best fit line horizontally.

- Move along this tick line to the left until you reach the vertical axis.

- Read off the value from the tick mark on the vertical axis.

Performing the above steps, the tick mark that you run into on the vertical axis is 3.3 inches. Notice that on the vertical axis each tick mark represents 0.1 inches. So, counting down two tick marks from 3.5 inches results in 3.3 inches.

The correct answer is A, 3.3 inches.

Problem 5

A curious high school student randomly selected 50 athletes from the list of all the athletes at her school. She asked each of the 50 athletes, "How many minutes per day do you typically spend studying?" The mean studying time in the sample was 52 minutes and the margin of error for this estimate was 5.64 minutes. Another high school student intends to replicate the survey and will attempt to get a similar margin of error. Which of the following samples will most likely result in a smaller margin of error for the estimated mean time that athletes study per day?

A) 15 randomly selected athletes

B) 15 randomly selected students from the entire high school.

C) 90 randomly selected athletes

D) 90 randomly selected students from the entire high school.

Solution and Explanation

A high school student is trying to recreate data already collected from a survey asking athletes how many minutes a day they study. In the new survey the high school student wants to reduce the margin of error. Before we get into the mathematical properties of the margin of error, please notice that the sample of students taken in the first survey consisted **only** of athletes. If the new survey were to include students from the entire student body in her survey it would make the new results rather unpredictable. So much so that you cannot predict how the margin of error will change because the sample of students has completely changed from athletes to anybody at the high school. Using this logic, we can already eliminate options B and D from our possible solutions.

Given that we now only need to consider options A and C for our possible solution, the question now becomes does the margin of error increase by choosing more or less students?

The margin of error, m, is inversely proportional to the sample size, n. Mathematically that is written as follows:

$$m \propto \frac{1}{n}$$

This is not the exact relationship between margin of error and sample size, rather it is a relationship of proportionality. That is why the proportionality symbol, α, was used in the place of an equal sign. This proportionality relationship will not give us an exact value for the margin of error, but it will tell us how the margin of error increases or decreases with increasing or decreasing values of the sample size.

Let's look at this proportionality relationship and see how the margin of error acts as we make the sample size larger. If $n = 1$, then the fraction $\frac{1}{n}$ becomes 1 and the margin of error is the same size as its actual value. If we let $n = 2$, then the margin of error becomes one half of its actual value. If we let $n = 3$, then the margin of error becomes one third of its actual value.

From these simple tests we can see that as the sample size becomes larger the margin of error becomes smaller. With this logic, looking back at the question, we can see that by increasing the sample size from 50 athletes to 90 athletes will decrease the margin of error.

The correct answer is D.

Problem 6

The table below shows a survey of 198 people from three different age groups. The survey asked them which genre of fiction they prefered.

	Teenagers	Adults	Elderly	Total
Science Fiction	34	52	12	98
Horror	18	23	5	46
Adventure	10	9	35	54
Total	62	84	52	198

Which fraction of all adults and teenagers prefered the adventure genre?

Solution and Explanation

There are some keywords to pay attention to in the question in order to answer it correctly. The first one is "fraction", this word gives away that we will be building a fraction. The words "of all" give away that whatever comes next, in this case adults and teenagers, will be in the denominator. And because they ask for all adults and teenagers, we need to find the sum of these two age groups. These numbers can be found in the Total row at the bottom of the Teenagers and Adults columns.

The total number of teenagers is 62 and the total number of adults is 84. The sum of these two numbers is $62 + 84 = 146$. Which means that 146 will go into the denominator of our fraction.

Since the denominator is filled, the only thing left to find is the number in the numerator. This will be the total number of adults and teenagers which prefered the adventure genre. There are 10 teenagers and 9 adults which prefered the adventure genre. The sum of these two numbers is $10 + 9 = 19$. Which means 19 will be in the numerator of our fraction.

The fraction and correct answer is $\dfrac{19}{146}$.

Problem 7

An irrigation company sent out 13 boxes with different amounts of pipe elbows in each box. The installation crew counted the number of elbows in each box and compiled the results in the form of a histogram as seen below.

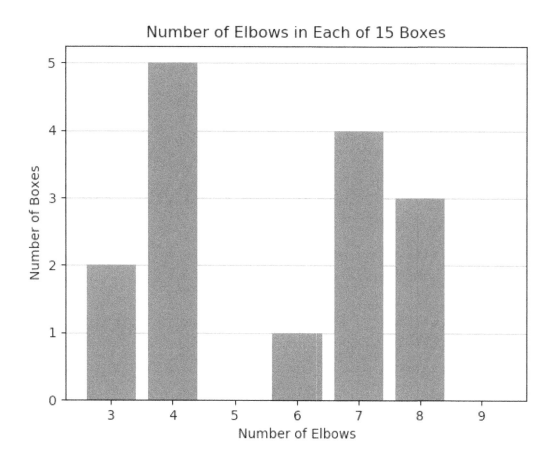

Of the following, which is closest to the average (arithmetic mean) number of elbows per box?

A) 3

B) 4

C) 5

D) 6

Solution and Explanation

This problem is asking you to find the average number of elbows per box. The histogram can be interpreted as follows:

- There are 2 boxes with 3 elbows.

- There are 5 boxes with 4 elbows.

- There are 0 boxes with 5 elbows.

- There is 1 box with 6 elbows.

- There are 4 boxes with 7 elbows.

- There are 3 boxes with 8 elbows.

- There are 0 boxes with 9 elbows.

To build the average (arithmetic mean) you need to sum up the total number of elbows and divide it by the number of boxes. I will sum up the elbows in the same order as in the list above so that you can follow along. Remember, the total number of boxes is 15.

$$\text{Avg.} = \frac{2(3) + 5(4) + 0(5) + 1(6) + 4(7) + 3(8) + 0(9)}{15}$$

$$\text{Avg.} = \frac{84}{15}$$

$$\text{Avg.} = 5.6$$

The average number of pipes per box is 5.6. Which from the solutions is closest to 6.

The correct answer is D.

Chapter 6

Additional Topics in Math

6.1 Geometry

Problem 1

The designers of the pyramids carved one side of the pyramid on stone to help them with their calculations. The dimensions of the triangle are shown below.

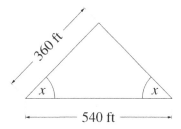

Note: Figure not drawn to scale.

What is the value of $\cos x$?

Solution and Explanation

The cosine of an angle equals the ratio of the adjacent side to the hypotenuse of a right triangle. In this problem; however, we do not have a right triangle to work with. But, because a similar angle x appears in both corners of this triangle, we can get two similar right triangles by drawing a line from the upper vertex straight down to the base of the triangle. This bold bisection line can be seen in the figure below, notice how this line is perpendicular with the base.

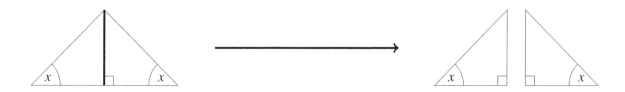

Either one of the right triangles on the right side of the figure above has a base with exactly half of the length of the original triangle base. Let's look at one of those triangles with its known dimensions in order to find $\cos x$.

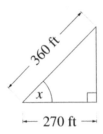

Now we can build the $\cos x$ simply by building the ratio of the adjacent side to the hypotenuse. Let's look at this relationship mathematically and solve the problem.

$$\cos x = \frac{\text{adjacent side length}}{\text{hypotenuse length}} = \frac{270 \text{ ft}}{360 \text{ ft}}$$

Now we have a ratio, but it still needs to be simplified. The first thing to do would be to see that the unit "ft" cancels out. Then we can rebuild the fraction in a factored-out form and cancel out like factors in the numerator and denominator.

$$\cos x = \frac{270}{360} = \frac{27 \times 10}{36 \times 10} = \frac{27}{36}$$

Still it can be simplified further.

$$\cos x = \frac{27}{36} = \frac{9 \times 3}{9 \times 4} = \frac{3}{4}$$

The correct answer is $\cos x = \dfrac{3}{4}$.

Problem 2

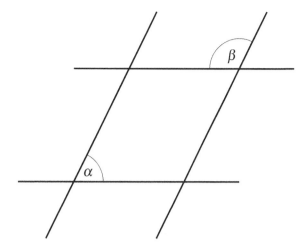

In the figure above, lines *a* and *b* are parallel and lines *y* and *z* are parallel. If the measure of α is 27°, what is the measure of β?

A) 63°

B) 27°

C) 153°

D) 70°

Solution and Explanation

There two important things to know when you see that two parallel lines intersect with two other parallel lines.

- The angle of a straight line is 180°.

- Only two angles exist between the intersecting lines. The sum of those two angles is equal to 180°.

It is one thing to know the rules above, but it will be much more effective memorization if you understand why this works. Given a straight line, it is a fact that the angle of that straight line is 180°.

If you now draw another straight line to intersect the line that we just drew, the sum of the angles between the intersecting lines will to be 180°.

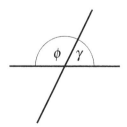

$$\phi + \gamma = 180°$$

Notice that the horizontal line was intersected by another straight line running nearly vertical on the page. It could also be said that the nearly vertical line was intersected by the horizontal line. In which case the angle below ϕ must be γ because the sum of the angles of an intersected straight line must equal 180° and $\phi + \gamma = 180°$.

The same reasoning can be used to say that the lower right angle must be equal to ϕ. In which case all the angles of the intersecting lines look as follows:

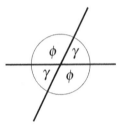

Now if we were to take the horizontal line, extend it out and allow another nearly vertical line, running parallel to the nearly vertical line above, run through and intersect with the horizontal line, it would be as if we made a copy of the nearly vertical line and shifted it over to the right. Because the nearly vertical lines are identical in direction, the angles they create when intersecting with the horizontal line must also be identical.

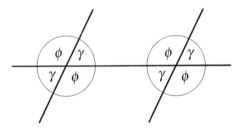

Before reading further, try on your own to extend the two nearly vertical lines out further and then run a horizontal line parallel to the one on the page through them. What will the angles of those intersecting lines look like? You can imagine taking the figure above, making a copy of it and placing it on top of the figure so that the nearly vertical lines match up. Doing that will provide a very clear picture of how to solve the SAT test problem. It looks as follows:

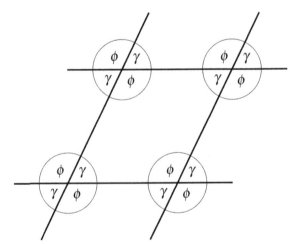

Comparing this picture with the one from the original problem we see that α is a γ and β is a ϕ. Given that we know the sum of γ and ϕ to be 180^{circ}, the sum of α and β must also be 180^{circ}. Plugging in the value of 27^{circ} for α we get the following equation:

$$\alpha + \beta = 27° + \beta = 180°$$

This equation can be solved for β by subtracting $27°$ from both sides of the equation.

$$27° - 27° + \beta = 180° - 27°$$
$$\beta = 153°$$

The correct answer is C.

Problem 3

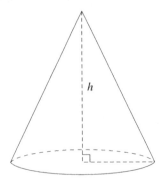

The volume of the cone above is 36 cubic inches and the circumference of the circle on the bottom of the cone is 4 inches. What is the height of the cone?

Solution and Explanation

This question is providing the volume of the entire cone and the circumference of the base of the cone. It asks you to find the height of the cone given that information. The equation for the volume of a cone will be provided for you when you take the SAT test, it is:

$$V = \frac{1}{3}\pi r^2 h$$

Where:

- V is the volume of the cone

- r is the radius of the circular base of the cone.

- h is the height of the cone.

The problem with the above equation is that there are two unknown variables, namely h and r. At this point don't forget that the circumference was also given. The equation for the circumference of a circle will also be provided to you on the SAT test, it is:

$$C = 2\pi r$$

Where:

- C is the circumference of the circle

- r is the radius of the circle

In the circumference equation there is only one unknown variable, r. And r is the variable we need to solve the volume equation.

To solve for the height of the cone, you first need to find the radius of its base with the circumference equation. To solve that equation for r divide the entire equation by 2π, in other words multiply the entire equation by $\frac{1}{2\pi}$

$$\frac{1}{2\pi}(C = 2\pi r)$$
$$\frac{C}{2\pi} = \frac{2\pi}{2\pi}r$$
$$\frac{C}{2\pi} = r$$

111

At this point it might be tempting to plug in the numerical value for C to solve for the numerical value of r. If you know you are going to solve for a numerical value only to plug that value into another equation, then my advice is don't bother. Keep the equation in variable form and plug those variables into the other equation. Speaking of which, take the equation for r and plug it into the volume equation.

$$V = \frac{1}{3}\pi r^2 h = \frac{1}{3}\pi \left(\frac{C}{2\pi}\right)^2 h$$

You can still simplify this equation further by distributing the square into the $\frac{C}{2\pi}$ expression. That will look as follows:

$$V = \frac{1}{3}\pi \left(\frac{C}{2\pi}\right)^2 h = \frac{1}{3}\pi \frac{C^2}{2^2\pi^2} h$$

Notice that the r equation brought in a π^2 into the denominator of the volume equation. One of the π cancels out with the π in the numerator, which leaves the following volume equation:

$$V = \frac{1}{3}\pi \frac{C^2}{2^2\pi^2} h = \frac{1}{3}\frac{C^2}{4\pi} h$$

Notice the 3 and 4 in the denominator of the equation? You can multiply those together to get a 12 in the denominator, so that the simplified form of the volume equation will look as follows:

$$V = \frac{C^2}{12\pi} h$$

Doing all of those simplifications really reduces the equation down to something that will be very easy to enter into your calculator after you have solved for h. It is often the case with geometrical equations that you can simplify them a lot and I would recommend practicing the simplifications from this problem a few times over the next couple weeks so that you can develop a stronger mathematical sense of simplifying equations.

To solve for h first multiply the entire equation by 12π, this will get rid of the 12π in the denominator of the right side of the equation.

$$12\pi \left(V = \frac{C^2}{12\pi} h\right)$$
$$12\pi V = \frac{12\pi}{12\pi} C^2 h$$
$$12\pi V = C^2 h$$

The final step to solve for h is to divide the entire equation by C^2. in other words multiply the entire equation by $\frac{1}{C^2}$.

$$\frac{1}{C^2}\left(12\pi V = C^2 h\right)$$
$$\frac{12\pi}{C^2} V = \frac{C^2}{C^2} h$$
$$\frac{12\pi}{C^2} V = h$$

With the equation completely solved for h plug in the numerical values $C = 4$ and $V = 36$ to find the numerical value for the height of the cone, h.

$$h = \frac{12\pi}{C^2} V = \frac{12\pi}{4^2} 36 = 84.78$$

$$h = 84.78$$

The height of the cone is 84.78 inches.

6.2 Trigonometry

Problem 1

In a right triangle, one angle measures ϕ degrees, where $\sin(\beta) = \frac{2}{3}$. What is the value of $\cos(-90° + \beta)$?

Solution and Explanation

In general, a right triangle looks as follows:

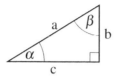

Where:

- a is the hypotenuse, the longest side of a right triangle

- b is the side opposite to the angle α and the side adjacent to the angle β

- c is the side opposite to the angle β and the side adjacent to the angle α

- the square on the bottom right corner of the triangle denotes that the angle is equal to $90°$

It is important to know that there are many relationships between the sine and the cosine functions. Often, due to the properties of the right triangle, these two functions can be set equal to each other; however, with different angles inside the parenthesis of the function. Let's look at one of those relationships as it relates to this problem.

$$\cos(\alpha) = \frac{\text{adjacent side length}}{\text{hypotenuse side length}} = \frac{c}{a}$$
$$\sin(\beta) = \frac{\text{opposite side length}}{\text{hypotenuse side length}} = \frac{c}{a}$$

This shows that both $\cos(\alpha)$ and $\sin(\beta)$ are equal to $\frac{c}{a}$. Because they are both equal to $\frac{c}{a}$, that means they are equal to each other. Look at it mathematically so that it is clearer:

$$\cos(\alpha) = \frac{c}{a} = \sin(\beta)$$
$$\cos(\alpha) = \sin(\beta)$$

Now that we have an idea of the relationship between the sine and cosine functions. We need to find a relationship between the angles, so that we can use the same angle within the parenthesis of both functions.

The sum of all the angles in a triangle is equal to $180°$. Which leads to the following equation:

$$\alpha + \beta + 90° = 180°$$

Because the sine function in the problem has the angle all by itself in the parenthesis, we are going to solve for α in the equation above. Doing this will leave the angle within our general equation for the sine function alone, giving us a better representation of the problem.

In order to solve for α you need to subtract β and $90°$ from both sides of the above equation.

$$\alpha + \beta - \beta + 90° - 90° = 180° - 90° - \alpha$$
$$\alpha = 180° - 90° - \beta$$
$$\alpha = 90° - \beta$$

Plugging this value in for α in the equation $\cos(\alpha) = \sin(\beta)$ above results in the following relationship between the sine and the cosine functions:

$$\cos(90° - \beta) = \sin(\beta)$$

The value inside the parenthesis of the cosine function is close to, yet not exactly the same as the value in the parenthesis of the cosine function from the problem. If you look closely you will notice that they differ by the value of negative 1. Let's manipulate the value inside of the parenthesis of the cosine functions so we can get the value from the problem multiplied by a negative 1. To do this factor out a negative 1 from the value within the parenthesis.

$$\cos(90° - \beta) = \cos(-1(-90° + \beta))$$

If this concept is new to you, go ahead and distribute the negative 1 back into the sum in the parenthesis to see that it is the same thing.

At this point you have almost solved the problem. You are being tested here on your knowledge of symmetrical functions. The cosine function is a symmetrical function, in that its curve is symmetrical about the vertical axis when plotted on a two-dimensional plane.

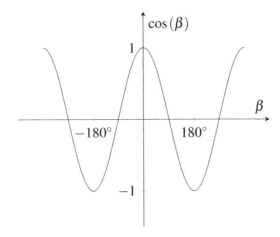

What the symmetry of the function means for this problem is that it doesn't matter if the value inside the parenthesis of the cosine function is positive β or negative β the result will be the same. Mathematically that looks as follows:

$$\cos(\beta) = \cos(-\beta)$$

Take a moment to look at the curve and prove to yourself that the above equation is in fact true for all values of β.

What that means for this problem is that the cosine function we derived above is in fact equal to the one from the problem.

$$\cos(90° - \beta) = \cos(-1(-90° + \beta)) = \cos(-90° + \beta)$$

Since we have already proved that $\sin(\beta) = \cos(90 - \beta)$ and given that the problem stated that $\sin(\beta) = \frac{2}{3}$, then the following must be true:

$$\sin(\beta) = \cos(90 - \beta) = \cos(-90° + \beta) = \frac{2}{3}$$

$$\cos(-90° + \beta) = \frac{2}{3}$$

The correct answer is $\frac{2}{3}$.

Chapter 7

Practice Test 1

Practice Math Test - No Calculators Allowed

Directions

You will have 25 minutes to complete part 3 of the math practice test. There are 20 questions for this portion of the test. Questions 1-15 are multiple choice and questions 16-20 are bubble-in questions where no choices are provided for you. You may use a piece of scratch paper for notes and calculations. Write your answers down on a separate piece of paper, the solutions and explanations can be found immediately after the test. You are to continue when you see the word CONTINUE in the bottom right corner of the page. Stop taking the test when the word STOP appears in the bottom center of the page. If you finish before the time expires you may check your answers for part 3 of the test only, do not move ahead within the 25 minutes provided for the portion of the test.

Notes

- No calculator usage allowed

- All variables and expressions belong to the set of real numbers unless otherwise written.

- Not all figures are drawn to scale. If the scaling is important it will be noted.

- If a drawing is in three-dimensions it will be noted, otherwise all drawings are in two-dimensions.

CONTINUE

Reference Equations

- The area of a circle is $A = \pi r^2$

- The circumference of a circle is $C = 2\pi r$

    r is the radius of the circle

- The area of a rectangle is $A = lw$

    l is the length of the rectangle

    w is the width of the rectangle

- The area of a triangle is $A = \dfrac{1}{2}bh$

    b is the base of the triangle

    h is the height of the triangle

- The sides of a right triangle have the following relationship: $a^2 + b^2 = c^2$

    c is the length of the hypotenuse

    a and b are the side lengths for the sides whose ends meet at the right angle

- A right triangle with angles of 30°, 60° and 90° angles has side lengths of x and $x\sqrt{3}$ and a hypotenuse length of $2x$.

- A right triangle with angles of 45°, 45° and 90° angles has side lengths of x and x and a hypotenuse length of $2x\sqrt{2}$.

- The volume of a rectangular prism is $V = lwh$

    h is the height of the prism

- The volume of a cylinder is $V = \pi r^2 h$

- The volume of a sphere is $V = \dfrac{4}{3}\pi r^3$

- The volume of a cone is $V = \dfrac{1}{3}\pi r^2 h$

- The volume of a pyramid is $V = \dfrac{1}{3}lwh$

- There are 360° in the arc of a circle

- There are 2π radians in the arc of a circle

- The sum of all the angles in a triangle equals 180°

CONTINUE

1.

If $12y - 17 = 7$, what is the value of $5y + 4$?

A) 19

B) -6

C) 14

D) -1

2.

$$x + 2y = 4$$
$$3x - y = 5$$

Which of the following ordered pairs (x, y) satisfies the system of equations above?

A) $(-2, 1)$

B) $(1, 2)$

C) $(1, -2)$

D) $(2, 1)$

3.

Julie writes books after school to make some extra money. She calculates that she makes $b + 0.02w$ per book. Where b is the flat rate paid per book and w is the number of words in the book. What is the best way to understand the number 0.02 in this expression?

A) The price Julie gets paid per word that she writes

B) The price Julie gets paid per page that she writes

C) The number of words Julie writes everyday

D) The price Julie gets paid per book that she writes

4.

$$16x^6 + 24x^3y^3 + 9y^6$$

What is the simplified form of the expression above?

A) $(16x^3 + 9y^3)^2$

B) $(4x^3 + 3y^3)^2$

C) $(4x^6 + 3y^6)^2$

D) $(16x^6 + 9y^6)^2$

CONTINUE

5.

$$t + \sqrt{4c^2 + 9} = 0$$

In the above equation, if $c > 0$ and $t = 5$, what is the value of c?

A) 5

B) 4

C) 3

D) 2

6.

Two lines run parallel with one another, one line passes through the points (-1, -1) and (1,2). The other line passes through the point (0,3), at which value of x does this line cross the x-axis?

A) -3

B) -4

C) -1

D) -2

7.

$$\frac{s^{2b}}{s^{2a}} = \frac{1}{s^4}$$

In the equation above, if $s > 0$, what is the value of $a - b$?

A) $\frac{1}{2}$

B) 1

C) 4

D) 2

8.

$$NV = 100$$

The number of marbles, N, with the volume, V, which can fit into a bucket with the volume of 100 cubic inches is described by the equation above. If the volume of a single marble is greater than 3 cubic inches, what is the maximum number of marbles that can fit in the bucket?

A) 33

B) 32

C) 34

D) 31

CONTINUE

9.

$$y = 2x - 4$$

$$y = \frac{2}{3}x$$

The system of equations above intersect at the point (a, b), what is the value of ab?

A) 5

B) 6

C) $\dfrac{2}{3}$

D) 4

10.

Which of the following equations has a graph in the xy-plane for which all values of x are less than or equal to 4?

A) $x = -y^2 + 5$

B) $x = -(y^2 + 5)$

C) $x = -y^3 + 5$

D) $x = -|y| + 5$

11.

What is the result of the multiplication of $3 + 2i$ with its complex conjugate? (Note: $i = \sqrt{-1}$)

A) $9 - 4i$

B) 13

C) $3 - 2i$

D) $13 + 12i$

12.

$$T = \frac{2P}{ql + P}$$

The above equation represents the tension found in a rope attached to a pulley. Where P is the force acting on the pulley, q is the force acting along the pulley's axle and l is the length of the axle. Which of the following equations expresses the force acting on the pulley in terms of the other variables?

A) $P = -\dfrac{Tql}{T - 2}$

B) $P = -\dfrac{Tql}{T + 2}$

C) $P = -\dfrac{Tql}{2 - T}$

D) $P = \dfrac{Tql}{T - 2}$

CONTINUE

13.

$$2x^2 - 18x + 18 = 0$$

What is the sum of all the values x that satisfies the above equation?

A) $-9 + 3\sqrt{5}$

B) $9 - 3\sqrt{5}$

C) -9

D) 9

14.

The production of go-karts is decreasing at a monthly rate of 2 percent. If the number of go-karts produced in the first year of this month was 150, which of the following functions g models the number of go-karts produced after m months?

A) $g(m) = 150(0.02)^m$

B) $g(m) = 150(0.98)^m$

C) $g(m) = 0.98(150)^m$

D) $g(m) = 0.02(150)^m$

15.

$$\frac{3t + 1}{t - 2}$$

Which of the following expressions is equivalent to the expression above?

A) $\dfrac{3 + 1}{-2}$

B) $3 - \dfrac{1}{2}$

C) $3 + \dfrac{1}{t - 2}$

D) $3 + \dfrac{7}{t - 2}$

CONTINUE

122

Directions

The final questions in part 3 of the test are to be bubbled in. On the actual test there will be pictures to show you how to properly bubble in the answers on your answer sheet. For this book, you only need to write down your answers on a separate sheet of paper. You will be able to bubble in fractions and decimals on the actual test, so don't worry if answers appear in this form on the practice test.

CONTINUE

16.

A local farmer gave away 250 apples to the local elementary school for making apple cider. He gave the apples away in full baskets containing either 25 or 50 apples. If at least two of the baskets were full with 25 apples and two of the baskets were full with 50 apples, what is one possible number of baskets with 25 apples brought to the elementary school?

17.

$$4s(3s - 2) + 2(5s + 1) = as^2 + bs + c$$

In the equations above, a, b, and c are constants. If the above equation is true for all values of t, what is the value of b?

18.

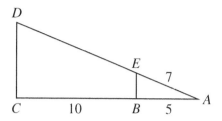

In the figure above $\overline{CD} \parallel \overline{BE}$, what is the length of the segment $\overline{DE}$?

CONTINUE

19.

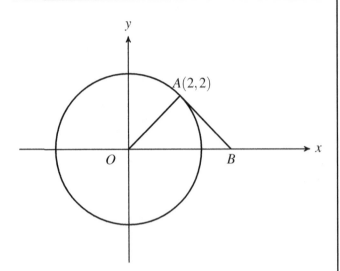

In the xy-plane above, O is the center of the circle, the line segment $\overline{AB}$ is tangent to the circle, the line segment $\overline{OA}$ is perpendicular to $\overline{AB}$ and the measure of $\angle ABO$ is $\dfrac{\pi}{b}$ radians. What is the value of b?

20.

$$ax + 3y = 17$$
$$2x - by = 68$$

In the system of equations above a and b are both constants. If the system of equations has an infinite number of solutions, what is the value of $a + b$?

STOP

If you finished early, check only your work for this section.
Do not look at any other section.

Directions

You will have 55 minutes to complete part 4 of the math practice test. There are 38 question in this portion of the test. Questions 1-30 are multiple choice and questions 31-38 are bubble-in questions, where no choices are provided for you. You may use a piece of scratch paper for notes and calculations. Write your answers down on a separate piece of paper, the solutions and explanations can be found immediately after the test. You are to continue when you see the word CONTINUE in the bottom right corner of the page. Stop taking the test when the word STOP appears in the bottom center of the page. If you finish before time expires you may check your answers for part 4 of the test only, do not return to part 3 within the 55 minutes provided for the portion of the test.

Notes

- Calculator usage is allowed

- All variables and expressions belong to the set of real numbers unless otherwise written.

- Not all figures are drawn to scale. If the scaling is important it will be noted.

- If a drawing is in three-dimensions it will be noted, otherwise all drawings are in two-dimensions.

CONTINUE

Reference Equations

- The area of a circle is $A = \pi r^2$

- The circumference of a circle is $C = 2\pi r$

    r is the radius of the circle

- The area of a rectangle is $A = lw$

    l is the length of the rectangle

    w is the width of the rectangle

- The area of a triangle is $A = \frac{1}{2}bh$

    b is the base of the triangle

    h is the height of the triangle

- The sides of a right triangle have the following relationship: $a^2 + b^2 = c^2$

    c is the length of the hypotenuse

    a and b are the side lengths for the sides whose ends meet at the right angle

- A right triangle with angles of $30°$, $60°$ and $90°$ angles has side lengths of x and $x\sqrt{3}$ and a hypotenuse length of $2x$.

- A right triangle with angles of $45°$, $45°$ and $90°$ angles has side lengths of x and x and a hypotenuse length of $2x\sqrt{2}$.

- The volume of a rectangular prism is $V = lwh$

    h is the height of the prism

- The volume of a cylinder is $V = \pi r^2 h$

- The volume of a sphere is $V = \frac{4}{3}\pi r^3$

- The volume of a cone is $V = \frac{1}{3}\pi r^2 h$

- The volume of a pyramid is $V = \frac{1}{3}lwh$

- There are $360°$ in the arc of a circle

- There are 2π radians in the arc of a circle

- The sum of all the angles in a triangle equals $180°$

CONTINUE

1.

A circus just came into town and is offering two different ticket prices for adults and children. The circus gets a profit of $3.00 for every adult and a profit of $7.00 for every child who purchases a ticket. Which of the following expressions represents the profit, in dollars, that the circus gets for every adult, A, and child, c, who attend the circus?

A) $3.00A + 7.00c$

B) $3.00A - 7.00c$

C) $3.00c + 7.00A$

D) $3.00c - 7.00A$

2.

A camel walking through the desert will step in quick sand 5 times every 120 miles. At this rate, how many times can you expect a camel to step in quick sand if it walks 2,280 miles?

A) 85

B) 90

C) 95

D) 100

3.

$$h = 1.9b + 32$$

The above equation is known to predict a person's height, h, given the length of the femur bone, b; all lengths are in inches. What is the femur length of a person who is 72 inches tall?

A) 21.1

B) 33.9

C) 72

D) 105.9

CONTINUE

128

The following information is for questions 4 and 5.

The amount of alley cats who appear outside of the local restaurant is directly proportional to the amount of spaghetti the owner leaves outside the backdoor. The restaurant owner counts 20 cats when he leaves 2 gallons of spaghetti by the backdoor.

4.

How many cats will appear by the restaurant's backdoor when the owner leaves 4.5 gallons out?

A) 54

B) 32

C) 45

D) 20

5.

Of the cats who show up at the restaurant 5% are taken directly to the animal shelter by pedestrians. How many cats are not taken to the animal shelter when the owner leaves 2 gallons of spaghetti out?

A) 15

B) 5

C) 19

D) 1

6.

If 15 is taken from t divided by 3 and 5 remains. How much will remain when t is doubled and 100 is taken from it?

A) 30

B) 20

C) 60

D) 15

7.

$$s = t^2 + 2t - 3$$

The equations above is a parabola in the ts-plane. Which of the following equations is equivalent to the one above and shows the t-intercepts?

A) $s + 3 = t^2 + 2t$

B) $s = (t^2 + 2)(t - 3)$

C) $s = (t + 3)(t - 1)$

D) $s = t^2 + 3(2t - 1)$

CONTINUE

8.

A farmer began the season with S functioning sprinklers. Every time a fox sneaks on to the farm the farmer loses the use of 5 sprinklers. If 23 foxes enter the farm throughout the season and the farm has 1,200 functioning sprinklers at the end of that time, then how many functioning sprinklers did the farmer have at the beginning of the season?

A) 1,085

B) 1,315

C) 1,200

D) 23

9.

A reservoir with 100,000 gallons of water is needed to supply two vineyards with water. Vineyard one is 36 acres and needs x gallons of water per acre. Vineyard two is 45 acres and needs y gallons of water per acre. Per minute, the total water supply to both fields must exceed 23 gallons of water per acre. Which system of inequalities best describes the water requirements for these two vineyards?

A) $36x + 45y \geq 100,000$
$x + y < 23$

B) $36x + 45y \leq 100,000$
$x + y > 23$

C) $\dfrac{x}{36} + \dfrac{y}{45} \geq 100,000$
$x + y < 23$

D) $\dfrac{x}{36} + \dfrac{y}{45} \leq 100,000$
$x + y > 23$

10.

If $f(x) = \dfrac{4}{2} + g(x)$ and $g(x) = 2x$, what is the value of $f(x)$ when $x = 2$?

A) 2

B) 3

C) 6

D) 7

11.

Number of hours jackhammer can run per day	4
Number of square feet jackhammer can crush per hour	12
Total area of blacktop at middle school	288
Total area of grass field next to blacktop	576
Number of children attending the middle school	1,080

A construction company is looking to tear apart the blacktop at a local middle school. The table above shows the capabilities of the company to destroy concrete with a jackhammer, and the size of the blacktop to be taken out. If the company works at the rates given in the table, which of the following is closest to the number of days it will take the construction company to destroy the entire blacktop with a jackhammer.

A) 4

B) 6

C) 12

D) 288

CONTINUE

12.

Online drip emitters at the top of a hill require a maximum of 18 PSI for proper operation. The pressure at the bottom of the hill is 72 PSI. As the water moves through the pipeline up the hill it loses 1 PSI for every 2.31 feet of elevation. With the elevation, z, in feet, which of the inequalities below describes the elevations for which the pressure is too high for proper operation of the drip emitters?

A) $\dfrac{1}{2.31}z > 18$

B) $72 + 18 \le \dfrac{1}{2.31}z$

C) $72z < 18$

D) $72 - \dfrac{1}{2.31}z > 18$

13.

A high school student randomly selected 3 students from her school and asked them whether they preferred reading at home or in the library. Everyone surveyed responded that they preferred to read in the library. Which of the following statistical factors takes away from the credibility of the results of the high school student?

A) Population size

B) Response rate

C) Sample size

D) Where the survey was conducted

14.

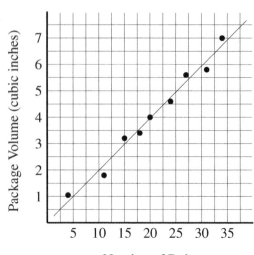

A warehouse manager created a scatter plot out of data comparing the number of bolts in a package to the package volume in cubic inches. How many bolts can the manager expect to find in a package which has a volume of 6.5 cubic inches?

A) 33

B) 27

C) 40

D) 13

CONTINUE

15.

Freddy rode his bike for 244 days in 2012 with an average of 16 hours a day. In the same year Freddy was able to travel 58,560 miles. What was Freddy speed in miles per minute?

A) 15

B) 0.25

C) 4

D) 61

16.

	Students who read the textbook	Students who did not read the textbook
Students who got an A	12	3
Students who did not get an A	1	25

The above table shows the results from a class of 41 students who were together in a 10th grade geography class. If one of the students, who did not get an A, stayed after class to talk with the teacher, what is the probability that the student read the textbook?

A) $\dfrac{25}{26}$

B) $\dfrac{12}{15}$

C) $\dfrac{3}{15}$

D) $\dfrac{1}{26}$

17.

The cost of a new bridge is estimated to be 15% less than that of a bridge built 15 years prior. If the bridge built 15 years ago cost $125,000, what is the estimated price of the new bridge?

A) $106,250

B) $18,750

C) $143,750

D) $231,250

18.

A survey of factories around country X showed that the mean production of washers per day per factory was 1,500,000 and the median production of washers per day per factory was 2,000,000. Which of the following scenarios would explain the difference between the mean and median washer production per day per factory?

A) All the factories in country X produce between 1,500,000 and 2,000,000 washers per day.

B) There are a greater number of factories which produced over 2,000,000 washers per day than those which produce less than 1,500,000 washers per day.

C) There are a greater number of factories that produce less than 1,500,000 washers per day than those which produce over 2,000,000 washers per day.

D) All the factories in country X produce the same number of washers per day.

CONTINUE

132

The following data is for questions 19 and 20.

19.

A marketing agent randomly chose 250 students from College X and College Y and asked them how many electronic devices they owned. The results can be found in the table below.

Number of Electronic Devices	College X	College Y
1	25	35
2	40	30
3	60	75
4	85	70
5	40	40

The total population of College X is 5,000 students. The total population of College Y is 6,000 students.

Of the 250 students surveyed at College X, what is the mean number of electronic devices owned? Round your answer to the nearest whole number.

A) 2

B) 3

C) 4

D) 5

The following question is based on data from the table in question 19.

20.

How many students from the entire population of both College X and College Y would you expect to own 4 electronic devices?

A) 155

B) 80

C) 3380

D) 3000

21.

A student estimates they will need t days to study thoroughly for the SAT test, where $t = 120$. They think the actual time needed, a, will be within 5 days of the estimate. Which of the following inequalities best describe the time frame the student has planned for studying?

A) $-5 \leq a - t \leq 5$

B) $a \geq t + 120$

C) $a \geq t + 5$

D) $a \leq t - 5$

CONTINUE

The following data is for questions 22 and 23.

22.

$$F = \frac{GmM}{r^2}$$

The above equation was developed by Sir Isaac Newton to describe the gravitational force, F, between two bodies with masses m and M separated by a distance r. The constant G is a gravitational constant.

Which of the following equations expresses the gravitational force, gravitational constant and the masses of the two bodies in term of the distance between the two bodies squared?

A) $r = \sqrt{\dfrac{GmM}{F}}$

B) $r = \sqrt{\dfrac{F}{GmM}}$

C) $r^2 = \dfrac{F}{GmM}$

D) $r^2 = \dfrac{GmM}{F}$

The following question refers to the equation found in question 22.

23.

Given that Newton's gravitational constant equals 6.674×10^{-11}. What is the gravitational force, F, between two masses, $m = 6 \times 10^7$ kilograms and $M = 5 \times 10^5$ kilograms, separated by 100 meters?

A) 2

B) 0.2

C) 20

D) 200

24.

Which of the following is an equation for a circle with its center located at (-3,-2) on the xy-plane and a diameter of 10?

A) $-\dfrac{x^2}{3} - \dfrac{y^2}{2} = 100$

B) $\dfrac{x^2}{3} + \dfrac{y^2}{2} = 25$

C) $(x+3)^2 + (y+2)^2 = 100$

D) $(x+3)^2 + (y+2)^2 = 25$

CONTINUE

25.

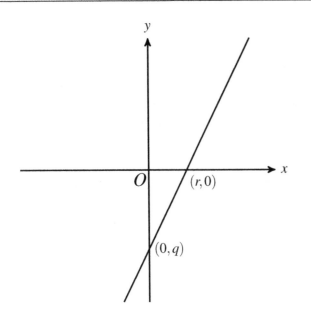

The line on the graph above intersects the *x* and *y* axes as shown above. If $r = 1$ and $r - q = 3$, what is the slope of a line perpendicular to the one above?

A) $-\dfrac{1}{2}$

B) $\dfrac{1}{2}$

C) 2

D) -2

26.

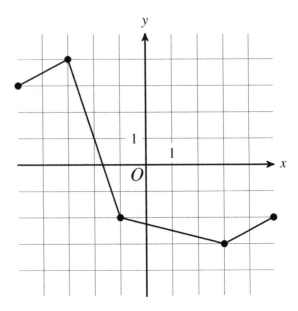

The function *g* is depicted in the graph above. Which of the following function values for *g* are less then -2?

I. $g(-\dfrac{5}{2})$

II. $g(0)$

III. $g(3)$

_____

A) I only

B) II only

C) I and III only

D) II and III only

CONTINUE

27.

Temperature Across Walls of Two Different Materials

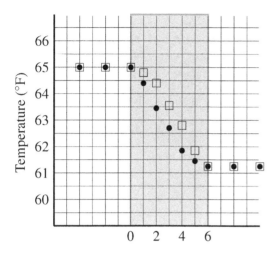

Position In The Wall (inches)

The squares are data points for material A. The circles are data points for material B

A temperature gradient is the measure of the change in temperature over a change in location. The graphic above shows temperature measurements at equal 1-inch intervals within two walls made of different materials. The temperature on both sides of the walls were held at constant temperatures of 65°F and 61.25°F. Which of the following statements best compares the average temperature gradients of the two different materials?

A) For every inch of material A the magnitude of the temperature gradient is greater than that of material B.

B) For every inch of material B the magnitude of the temperature gradient is greater than that of material A.

C) From 1 to 2 inches the magnitude of the temperature gradient for material B is greater than that of material A. From 4 to 5 inches the magnitude of the temperature gradient for material A is greater than that of material B.

D) From 1 to 2 inches the magnitude of the temperature gradient for material A is greater than that of material B. From 4 to 5 inches the magnitude of the temperature gradient for material B is greater than that of material A.

28.

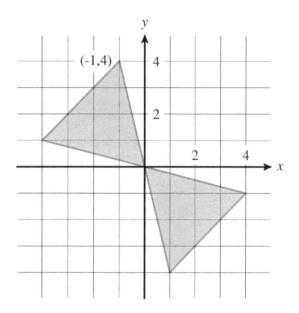

Two isosceles triangles mirror each other on the xy-plane above. The length of the bases of the triangles equals $\sqrt{18}$. Which of the following is the area of both of the triangles shown on the graph above?

A) 53

B) 26.5

C) 13.3

D) 106

CONTINUE

136

29.

$$y = -1$$
$$(x-a)^2 + (y-b)^2 = 9$$

The systems of equations above contains the constants a and b. For which of the following values of the constants (a,b), does the above system of equations have exactly one solution?

A) (0,1)

B) (0,0)

C) (0,2)

D) (0,3)

30.

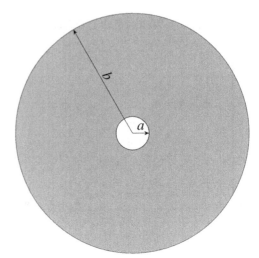

$a = 0.25$ inches, $b = 2$ inches

The compact disc above has the measurements shown and a thickness of 0.0625 inches. If 780 compact discs were stacked on top of one another, what would the volume of the stack of CDs be?

Hint: The volume of the hole through the middle of the stack is not a part of the total volume.

A) 613

B) 603

C) 3

D) 244

CONTINUE

Directions

The final questions in part 4 of the test are to be bubbled in. On the actual test there will be pictures to show you how to properly bubble in the answers on your answer sheet. For this book, you only need to write down your answers on a separate sheet of paper. You will be able to bubble in fractions and decimals on the actual test, so don't worry if answers appear in this form on the practice test.

CONTINUE

31.

Timothy can swim freestyle at a rate of 100 meters per minute. If he swam non-stop at this rate, how many minutes would it take him to swim one kilometer?

32.

If z feet and 8 inches is 188 inches, what is the value of z?

33.

When the function $g(x) = 2x^3 - cx^2 + 3x - 14$ is graphed on the xy-plane it runs through the point (2,-12). What is the value of the constant c in this function?

34.

Last month Chuck and Sarah ate a combined 370 cheeseburgers. Sarah ate 70 more cheeseburgers than Chuck. How many cheeseburgers did Sarah eat?

CONTINUE

35.

$$c = 7m + 23$$

Stuart started a coin collection after his neighbor gave him some coins to start off with. After that he added a fixed amount of coins every month to his collection. The model above represents the amount of coins, c, that Stuart has m months after starting his collection. What is the initial amount of coins given to Stuart by his neighbor?

36.

In the graphic above a triangle lies on a straight line. The angles between the outside of the triangle and the straight line are given. What is the value of the angle A?

CONTINUE

The following information is for questions 37 and 38

37.

The Newton-Raphson Method below is an algorithm for solving algebraic equations.

$$x_n = x_{n-1} - \frac{f(x_{n-1})}{f'(x_{n-1})}$$

Where $f'(x_{n-1})$ is the first derivative of $f(x_{n-1})$.

_____

$$f(x_{n-1}) = 3x_{n-1}^3 + 2x_{n-1}^2 + 1$$
$$f'(x_{n-1}) = 9x_{n-1}^2 + 4x_{n-1}$$

Given the equations above and that $x_{n-1} = 2$, what is the value of x_n?

38.

If $f(x_{n-1}) = 5x_{n-1} + 4$, $x_{n-1} = 3$ and $x_n = 2$ what is the value of $f'(x_{n-1})$?

STOP
If you finished early, check only your work for this section.
Do not look at any other section.

7.1 Practice Test 1 Solutions

Part 3: Multiple Choice

1. C
2. D
3. A
4. B
5. D
6. D
7. D
8. A
9. B
10. B
11. B
12. A
13. D
14. B
15. D

Part 3: Bubble In

16. 2, 4 or 6
17. 2
18. 14
19. 4
20. -11.5

Part 4: Multiple Choice

1. A
2. C
3. A
4. C
5. C
6. B
7. C
8. B
9. B
10. C
11. C
12. D
13. C
14. A
15. B
16. D
17. A
18. B
19. B
20. C
21. A
22. D
23. B
24. D
25. C
26. D
27. C
28. A
29. C
30. B

Part 4: Bubble In

31. 10
32. 15
33. -5
34. 150
35. 23
36. 100
37. 1.25
38. 19

7.2 Practice Test 1 Explanations

Question 1

Move the 7 from the right side to the left side of the equation.

$$12y - 17 - 7 = 7 - 7$$
$$12y - 24 = 0$$

Multiply the entire equation by $\dfrac{5}{12}$ to turn the $12y$ into $5y$.

$$\frac{5}{12}(12y - 24 = 0)$$
$$5y - 10 = 0$$

Add 14 to both sides of the equation to turn the -10 into a 4.

$$5y - 10 + 14 = 0 + 14$$
$$5y + 4 = 14$$

The correct answer is C.

Question 2

Solve the bottom equation, $3x - y = 5$ for y.

$$y = 3x - 5 \tag{7.1}$$

Plug y into the top equation, $x + 2y = 4$.

$$x + 2(3x - 5) = 4$$

Distribute the 2, collect the x-terms on the left side of the equation and non x- terms on the right side of the equation.

$$x + 6x - 10 = 4$$
$$7x - 10 + 10 = 4 + 10$$
$$7x = 14$$

Solve by multiplying the entire equation by $\dfrac{1}{7}$.

$$\frac{1}{7}(7x = 14)$$
$$x = 2$$

Plug the value for x into the equation for y. That was $y = 3x - 5$, solve for y.

$$y = 3(2) - 5 = 6 - 5 = 1$$
$$y = 1$$

The coordinates are (2,1).

The correct answer is D.

Question 3

The variable b in the sum $b + 0.02w$ is dollars. Which means the product $0.02w$ must also be in dollars. The question says the units of w are words, so the units of 0.02 must be dollars per word so that the words can cancel out. If the units of 0.02 are dollars per word, then the answer must be A; the price Julie gets paid per word that she writes.

The correct answer is A.

Question 4

At first glance it might not be so obvious how to break down the polynomial in this question. Look at the answers, they are all sums squared. Which means the F.I.O.L must be implemented. Since the first term in each of the sums are different, you need only multiply the first term in each sum with itself to see if it will be a term in the polynomial from the question. The process looks as follows, starting from option A:

$$16x^3(16x^3) = 16(16)x^{3+3} = 256x^6 \text{ NOT IN POLYNOMIAL FROM QUESTION}$$
$$4x^3(4x^3) = 4(4)x^{3+3} = 16x^6 \text{ First Term in the Polynomial from Question}$$

Once you see that you have found a term in the polynomial from the question. Stop right there, for this question, you have done enough to answer it.

The correct answer is B.

Question 5

The value of t is given as 5. which means this question is asking you to solve for c. To get c by itself, the first thing you need to do is move the t or 5 to the right side of the equation.

$$t - t + \sqrt{4c^2 + 9} = 0 - t$$
$$\sqrt{4c^2 + 9} = -t$$

Now you need to square both sides of the equation to get rid of the square root.

$$(\sqrt{4c^2 + 9})^2 = (-t)^2$$

Replace the t with 5 and reduce the squared square root.

$$4c^2 + 9 = 25$$

Subtract 9 from both sides and then multiply the entire equation by $\dfrac{1}{4}$

$$4c^2 + 9 - 9 = 25 - 9$$
$$4c^2 = 16$$
$$\frac{1}{4}(4c^2 = 16)$$
$$c^2 = 4$$

In order to find the value of c you need to take the square root of both sides of the equation.

$$\sqrt{c^2} = \pm\sqrt{4}$$
$$c = \pm 2$$

The question says that $c > 0$, which means c must equal 2.

The correct answer is D.

Question 6

If two lines are running parallel to one another, then they have the same slope. The slope of the line with the two points given is the ratio of the difference in the y-positions to the difference in the x-positions.

$$\frac{-1-2}{-1-1} = \frac{-3}{-2} = \frac{3}{2}$$

Since the slope is rise over run, go to the point given for the second line in the problem $(0,3)$. Here, if you were to continue with the positive slope you will never run into the x-axis. So instead of rising by positive 3 and running by positive 2. You need to rise by negative 3, which will bring you to the x-axis, and run by negative 2, which will bring you to the point $(-2,0)$.

The correct answer is D.

Question 7

The equation

$$\frac{s^{2b}}{s^{2a}} = \frac{1}{s^4}$$

can be rewritten as

$$s^{2b-2a} = s^{-4}$$

Because both exponents are over the variable s, they must be equivalent. Which means the following must be true:

$$2b - 2a = -4$$

The question is looking for $a - b$, so in this scenario you want to divide the entire equation by -2. In other words, multiply the equation by $-\frac{1}{2}$.

$$-\frac{1}{2}(2b - 2a) = -4$$
$$-\frac{2}{2}b + \frac{2}{2}a = \frac{4}{2}$$
$$a - b = 2$$

The correct answer is D.

Question 8

Rewrite the equation $NV = 100$ as $V = \dfrac{100}{N}$ by dividing the entire equation by N.

In the question it is stated that the volume of a single marble is greater than 3. This needs to be interpreted as the presence of an inequality in the problem, namely $V > 3$. Take the equation $V = \dfrac{100}{N}$ and plug it into $V > 3$, which results in the following:

$$\frac{100}{N} > 3$$

The question is looking for the maximum number of marbles which can fit into the bucket. If you solve the above inequality it will provide the answer to the question. To solve the inequality multiply the entire inequality by N.

$$N(\frac{100}{N} > 3)$$
$$\frac{N}{N}100 > 3N$$
$$100 > 3N$$

Now divide the entire equation by 3, to see what the maximum value for N is.

$$\frac{100}{3} > N$$
$$33.3 > N$$

The above inequality says that the number of marbles needs to be less than 33.3. However, because the number of marbles needs to be a whole number, then the maximum number of marbles allowed is 33.

The correct answer is A.

Question 9

For this question notice that if you solve the bottom question for x, then the 2 will cancel out in the top equation. This is something that is easier to see with practice. Solving the bottom equation for x requires that you multiply the entire bottom equation by $\dfrac{3}{2}$.

$$\frac{3}{2}\left(y = \frac{2}{3}x\right)$$
$$\frac{3}{2}y = \frac{3}{2}\frac{2}{3}x$$
$$\frac{3}{2}y = x$$

Plug this value of x in terms of y into the top equation, that will look as follows:

$$y = 2\frac{3}{2}y - 4$$
$$y = 3y - 4$$

146

See how the 2 canceled out in the equation above? Now move the $3y$ to the left side of the equation by subtracting $3y$ from both sides.

$$y - 3y = 3y - 3y - 4$$
$$-2y = -4$$

Now multiple the entire equation by $-\frac{1}{2}$ so that you can get y all by itself.

$$-\frac{1}{2}(-2y = -4)$$
$$\frac{2}{2}y = \frac{4}{2}$$
$$y = 2$$

Plug this value for y into the equation $\frac{3}{2}y = x$. Here again the 2s will cancel out so that $x = 3$.

For this equation $a = x = 3$ and $b = y = 2$. The product ab equals 3 times 2, which equals 6.

The correct answer is B.

Question 10

This problem is looking for a solution which is always less than 4. Each of the solutions presented are parabolas and have maximums when $y = 0$.

For solution A, when $y = 0$, then $x = 5$. This value is greater than 4, so this cannot be the correct answer.

For solution B, when $y = 0$, then $x = -5$. As y increases, the value of x will become more and more negative, so this solution is correct.

For solution C, when $y = 0$ then $x = 5$. This also cannot be the correct solution.

For solution D, when $y = 0$ then $x = 5$. This also cannot be the correct solution.

The correct answer is B.

Question 11

The complex conjugate of an imaginary number results when the sign in front of the imaginary component is changed. For the imaginary number in this question, $z = 3 + 2i$, the complex conjugate $z^* = 3 - 2i$. The multiplication of this imaginary number with its complex conjugate looks as follows using the F.I.O.L. method:

$$(3 + 2i)(3 - 2i) = 9 + 6i - 6i - 4i^2$$

Notice that the inner terms will cancel out. Also remember that $i^2 = \sqrt{-1}^2 = -1$, which means the final solution will be calculated as follows:

$$9 - 4i^2 = 9 - 4(-1) = 9 + 4 = 13$$

The correct answer is B.

Question 12

This question is asking you to show off your algebra skills and rearrange an equation written in terms of T into an equation in terms of P. The first thing to do is to multiply the entire equation by the sum $ql + P$, so that the sum in the denominator is canceled out.

$$(ql + P)\left(T = \frac{2P}{ql + P}\right)$$
$$T(ql + P) = \frac{ql + P}{ql + P}2P$$
$$T(ql + P) = 2P$$

Distribute the variable T into the sum $ql + P$.

$$Tql + TP = 2P$$

In this case subtract TP from both sides of the equation, so that all P terms are on the right side and all non-P terms on the left side.

$$Tql + TP - TP = 2P - TP$$
$$Tql = 2P - TP$$

Now you need to factor the P out of the sum on the right side of the equation.

$$Tql = 2P - TP = P(2 - T)$$

To solve this equation exclusively for P divide the entire equation by $2 - T$, in other words multiply the entire equation by $\frac{1}{2 - T}$. Notice, however, that all of the solutions with $2 - T$ in the denominator have a negative sign in front, namely solution C. In this situation we need to factor out a negative 1 from the sum $2 - T$ so that you directly get to the solution. Factoring out a negative 1 from the sum will look as follows:

$$2 - T = -(T - 2)$$

Do the math on that and notice that when you multiply the sum on the right side of the equation it will result in $2 - T$.

Which means the equation above, $Tql = P(2 - T)$ will become $Tql = -P(T - 2)$. Now we are going to multiply the entire equation by $-\frac{1}{T - 2}$.

$$-\frac{1}{T - 2}(Tql = -P(T - 2))$$
$$-\frac{Tql}{T - 2} = \frac{-(T - 2)}{-(T - 2)}P$$
$$-\frac{Tql}{T - 2} = P$$

The correct answer is A.

Question 13

This question is asking for the sum of the solutions. Which means first we need to solve the quadratic equation and then add up the solutions. The first thing to do to this equation is to multiply the entire equation by $\frac{1}{2}$. This will eliminate the factor of 2 from the x^2 term, it is also practical because all the numbers are divisible by 2.

$$\frac{1}{2}(2x^2 - 18x + 18 = 0)$$
$$\frac{2}{2}x^2 - \frac{18}{2}x + \frac{18}{2} = \frac{0}{2}$$
$$x^2 - 9x + 9 = 0$$

This quadratic equation needs to be plugged into the quadratic formula because it cannot easily be factored. The quadratic formula is:

$$x = \frac{-b \pm \sqrt{b^2 - 4ac}}{2a}$$

The quadratic formula should be memorized. Here a is the factor in front of the x^2 term, b is the factor in front of the x term and c is the constant term. So that $a = 1$, $b = -9$ and $c = 9$. Which means the quadratic formula will look as follows:

$$x = \frac{-(-9) \pm \sqrt{(-9)^2 - 4(9)}}{2} = \frac{9 \pm \sqrt{81 - 36}}{2} = \frac{9 \pm \sqrt{81 - 36}}{2} = \frac{9 \pm \sqrt{45}}{2}$$

Here you need to know how to work with square roots. Namely, you need to be able to do the following:

$$\sqrt{45} = \sqrt{9(5)} = \sqrt{9}\sqrt{5} = 3\sqrt{5}$$

So that the quadratic formula we were working on above can be rewritten as:

$$x = \frac{9 \pm \sqrt{45}}{2}$$
$$x = \frac{9 \pm 3\sqrt{5}}{2}$$
$$x = \frac{9}{2} \pm \frac{3\sqrt{5}}{2}$$

Now, there are two solutions for x because of the $\pm$ sign. So, the two solutions for x which need to be added up are.

$$x = \frac{9}{2} + \frac{3\sqrt{5}}{2}$$
$$x = \frac{9}{2} - \frac{3\sqrt{5}}{2}$$

The sum of the two solutions will be as follows:

$$(\frac{9}{2} + \frac{3\sqrt{5}}{2}) + (\frac{9}{2} - \frac{3\sqrt{5}}{2}) = \frac{9}{2} + \frac{9}{2} = 9$$

Since the sum of the two solutions is 9, the correct answer is D.

Question 14

As per the question, the starting value is 150. This will decrease by 2% every month. In one month a 2% decrease is represented mathematically as 150(0.98). To get a decrease in the second month of 2% you need to multiply the first month's decreased value by 0.98, which can be written mathematically as 105(0.98)(0.98), which equals $105(0.98)^2$. After 3 months, $m = 3$, the initial value, decreased by 2% each month, will equal $105(0.98)^3 = 105(0.98)^m$.

The correct answer is B.

Question 15

Simplifying a fraction with a sum in the denominator requires you to factor out the same sum from terms in the numerator. This may sound tricky or complicated but watch and try to follow along.

How the fraction is written in the problem does not allow for cancellation of the sum in the denominator. In this case, if we were able to subtract 7 from the 1 then there would be $3t - 6$ in the numerator and you could factor out the 3 to get $3(t-2)$. This would be perfect, except that you can't just subtract 7 because you feel like it. If you subtract 7 you also need to add a 7 to cancel out the subtraction. Let's see how that looks mathematically:

$$\frac{3t - 7 + 7 + 1}{t - 2}$$

Again, you subtract the 7 because it makes life easy for you. And you add the 7 because a mathematical expression can only be changed by multiplication with 1 or by the addition of 0. The subtraction of 7 followed by the addition of 7, is the equivalent of adding 0, an operation which is allowed to be performed on a mathematical expression. Now, rearrange the expression above, so that you can factor out the sum $t - 2$ in the numerator.

$$\frac{3t - 7 + 7 + 1}{t - 2} = \frac{3t - 6 + 7}{t - 2} = \frac{3(t-2) + 7}{t - 2}$$

Once you have factored the 3 out of the numerator you need to split the fraction into two fractions. And then cancel out the $t - 2$ sum from the left most term in the sum.

$$\frac{3(t-2) + 7}{t - 2} = \frac{3(t-2)}{t - 2} + \frac{7}{t - 2} = 3 + \frac{7}{t - 2}$$

The correct answer is D.

Question 16

A farmer gave away 250 apples. From those 250 apples at least 50 were given away in full baskets of 25 apples and 100 apples were given away in full baskets that carried 50 apples. In other words, of the 250 apples given away, the method for which 150 of the apples were given away is already accounted for with the 4 baskets, 2 filled with 25 apples and 2 filled with 50 apples, leaving only 100 apples left for consideration.

With the 2 full baskets of 25 already considered, how many more baskets of 25 apples could have been used to deliver the 250 apples from the farmer.

- 4 baskets of 25 and 0 baskets of 50

- 2 baskets of 25 and 1 basket of 50

- 0 baskets of 25 and 2 baskets of 50

So, the different possible numbers of baskets used to deliver the apples to the elementary school can be 2, 4 or 6.

Question 17

The question is asking you to solve the coefficient of the s term from the quadratic equation on the left side of the equation in the question. Start by distributing anything possible on the left side of the equation into their respective sums. That looks as follows:

$$4s(3s-2)+2(5s+1)=12s^2-8s+10s+2$$

Now you need to combine the like terms, as follows:

$$12s^2-8s+10s+2=12s^2+2s+2$$

The as^2+bs+c form of the quadratic equation is written on the right side of the equation above. Comparing the coefficients shows that $b=2$.

Question 18

This is a problem of similar triangles. The ratio of the length of one side of a triangle to the same side of a similar triangle is equal to the ratio of any other side to its same side on the similar triangle. Again, that may have been a tongue-twister, but in terms of this problem it can be expressed more understandably as the ratio of 15 to 5 (15 because the length of the same side on the similar triangle is 10 + 5) is equal to the ratio of $(\overline{DE}+7)$ to 7. That can be written mathematically as follows:

$$\frac{15}{5}=\frac{\overline{DE}+7}{7}$$
$$3=\frac{\overline{DE}+7}{7}$$

To solve this equation for $\overline{DE}$, first multiply the entire equation by 7.

$$7(3=\frac{\overline{DE}+7}{7})$$
$$21=\frac{7}{7}\overline{DE}+7$$
$$21=\overline{DE}+7$$

Now to solve for $\overline{DE}$ subtract 7 from both sides of the equation.

$$21-7=\overline{DE}+7-7$$
$$14=\overline{DE}$$

The correct answer is 14.

Question 19

The triangle in this problem is a right triangle with its right angle at the point A. This right triangle can in turn be cut in half into two identical right triangles. This can be done by drawing a vertical line from the point A down to the base of the triangle.

The point A at (2,2) gives away the side lengths of the two identical right triangles. The two side lengths given away by this point also happen to be the two side lengths which determine the slope of the triangles.

You need to pay special attention to the right triangle of the two identical triangles, because that is where the angle $\angle ABO$ is, next to point B. The tangent of this angle equals the slope of the triangle, this can all be written mathematically as follows:

$$\tan \angle ABO = \tan \frac{\pi}{b} = \frac{2}{2}$$
$$\tan \frac{\pi}{b} = 1$$

This is calling on your basic knowledge of trigonometry. The test setters are expecting you to know that $\tan 45° = 1$, they are also expecting you to know that π radians equals $180°$. Since $45°$ is one fourth of $180°$. Then you are looking for a value one fourth that of π, which is $\frac{\pi}{4}$.

The answer to this question is $b = 4$.

Question 20

For the system of equations to have infinitely many solutions, the two equations need to be exactly the same. To find out what to multiply one of the equations by, take the number from the right side of the equation with a larger value on the right side, in this case 68, and divide it by the value on the right side of the equation, in this case 17. 68 divided by 17 is 4. This means that you need to multiply the top equation by 4, that will result in both equations being equivalent. Performing the multiplication will result in the following system of equations:

$$4ax + 12y = 68$$
$$2x - by = 68$$

Because there are infinitely many solutions to this system of equations, the coefficient of the x-terms must be equal and the coefficients of the y-terms must be the same. Which means the following equations are true:

$$4a = 2$$

and

$$12 = -b$$

Starting with the equation for a. To solve that equation multiply the entire equation by $\frac{1}{4}$.

$$\frac{1}{4}(4a = 2)$$
$$\frac{4}{4}a = \frac{2}{4}$$
$$a = \frac{1}{2} = 0.5$$

Solving for b in the second equation only requires that you multiply the entire equation by negative 1.

$$-1(12 = -b)$$
$$-12 = b$$

Given the values of a and b, the value $a + b$ is 0.5 - 12.5.

The solution is -11.5.

152

Question 1

To solve this problem multiple the profit per adult, $3.00, by the number of adults, A. Add that product to the multiplication of the profit per child, $7.00, with the number of children, c. That will result in the sum $3.00A + 7.00c$.

The correct answer is A.

Question 2

A camel will step in quicksand 5 times every 120 miles. How many times will a camel walk 120 miles if it walks 2,280 miles?

$$\frac{2,280}{120} = 19$$

If the camel walks 120 miles 19 times. The number of times it will step in quicksand is 19 times 5 which is 95.

The correct answer is C.

Question 3

This question is writing a formula for a person's height h in terms of their femur bone length b. The question is however asking for the femur length of a person, given the person's height. Which means you will need to solve the equation for h in terms of b. The first thing to do is to subtract 32 from both sides of the equation.

$$h - 32 = 1.9b + 32 - 32$$
$$h - 32 = 1.9b$$

Now, you need to get b all by itself. To do that multiply the entire equation by $\frac{1}{1.9}$. That looks as follows:

$$\frac{1}{1.9}(h - 32 = 1.9b)$$
$$\frac{h - 32}{1.9} = \frac{1.9}{1.9}b$$
$$\frac{h - 32}{1.9} = b$$

Now all that is left to do is plug in the height of 72 for h and solve for b.

$$b = \frac{72 - 32}{1.9} = \frac{40}{1.9} = 21.1$$

The correct answer is A.

Question 4

The amount of alley cats, call them c, is proportional to the amount of spaghetti, call that s. When two things are proportional, it is best to set up an equation with a linear proportionality constant. In this case the equation will look as follows:

$$c = ks$$

Where k is a constant of proportionality. The information given of 20 cats and 2 gallons of spaghetti will allow you to solve for the constant, k. First solve for k in the equation by dividing the entire equation by s, in other words multiply the equation by $\frac{1}{s}$

$$\frac{1}{s}(c = ks)$$
$$\frac{c}{s} = k$$

With this equation plug in 20 for c and 2 for s. There you will find that k is equal to 10.

Question 4 gives the value of the spaghetti, s, as 4.5. You know from the original proportionality equation that if you multiply 4.5 by 10, the solution for the number of cats is 45.

The correct answer is C.

Question 5

When 2 gallons of spaghetti are left out, 20 cats will be at the restaurant. Of those 20 cats, 5% will be taken to the animal shelter. 5% is one twentieth, which means 1 out of those 20 cats will be taken to the shelter. 19 of those cats will not be taken to the shelter.

The correct answer is C.

Question 6

15 taken from t divided by 3, write that down mathematically as follows:

$$\frac{t}{3} - 15$$

From that expression 5 will remain. That means it needs to be set equal to 5.

$$\frac{t}{3} - 15 = 5$$

The next piece of information is asking you to double t, in order to do this, you will first need to solve for the value of t. Use the equation above and add 15 to both sides.

$$\frac{t}{3} - 15 + 15 = 5 + 15$$
$$\frac{t}{3} = 20$$

To get t all by itself you will need to multiply the entire equation by 3.

$$3(\frac{t}{3} = 20)$$
$$t = 60$$

Double t will result in 120. Taking 100 from that value will result in 20.

The correct answer is B.

Question 7

The only equation which shows t-intercepts is shown in answer C. It can also be called the factored form of the polynomial

The correct answer is C.

Question 8

Multiply the number of foxes who entered the farm throughout the season by the number of sprinklers destroyed per visit paid by the foxes. That is 23 times 5, which equals 115. That number added to the number of sprinklers still functioning at the end of the season, 1,200, will result in the number of functioning sprinklers at the beginning of the season.

$$1,200 + 115 = 1,315$$

The correct answer is B.

Question 9

The reservoir only has 100,000 gallons of water. The water supplied to each field is equal to the size of the field in acres, multiplied by the water supplied to the field per acre. That value needs to be less than or equal to 100,000. Mathematically the following information can be expressed in equation form as:

$$36x + 45y \leq 100,000$$

The water supplied per acre needs to be greater than 23 gallons per acre.

$$x + y > 23$$

Both of these equations are to be found in answer B.

The correct answer is B.

Question 10

Plug the function for $g(x)$ into the function for $f(x)$ so that $f(x)$ can be written as:

$$f(x) = \frac{4}{2} + 2x = 2 + 2x$$

When $x = 2$ the function produces the following value:

$$f(x = 2) = 2 + 2(2) = 2 + 4 = 6$$

The correct answer is C.

Question 11

The jack hammer can run 4 hours a day and destroy 12 square feet of the black top per hour. This all means that per day the jack hammer can destroy 4 times 12 square feet or 48 square feet of blacktop per day. Divide the entire surface area of the blacktop by the number of square feet which the jack hammer can take on in a day and you will get the number of days it will take for the jack hammer to take apart the entire blacktop.

$$\frac{288}{48} = 6$$

The correct answer is C.

Question 12

The water at the bottom of the hill has a pressure of 72 PSI. For every 2.31 feet of elevation, the pressure will drop by 1 PSI. For every foot increase in elevation, the pressure in this particular pipeline can be expressed by the following function in terms of the elevation z.

$$72 - \frac{1}{2.31}z$$

For all elevations, z, for which the pressure is above 18 PSI, the emitters will not be operational.

$$72 - \frac{1}{2.31}z > 18$$

The correct answer is D.

Question 13

The problem with these statistics is the sample size. A good rule of thumb is to have 30 samples in order to get good statistics.

The correct answer is C.

Question 14

The question is asking for the number of bolts which can be expected to be found in a box with a volume of 6.5 cubic inches. Since the cubic inches are given look along the vertical axis until you come to 6.5. Then go horizontally to the right until you run into the best fit line. Once you hit the best fit line, go vertically down and read off the value from the horizontal axis. That value is 32.5 bolts, which is closest to 33.

The correct answer is A.

Question 15

Freddy rode his bike on 244 days for 16 hours a day, which is a total of $244(16) = 3,904$ hours. That is a total of:

$$3,904 \text{ hours} \left(\frac{60 \text{ minutes}}{\text{hour}} \right) = 234,240 \text{ minutes}$$

Take the total number of miles traveled, 58,560, and divide it by the total number of minutes to get the average speed in miles per minute.

$$\frac{58,560}{234,240} = 0.25$$

The correct answer is B.

Question 16

The pool to choose from in this question is the students who did not get an A in the class. There were 26 of those students. The probability that the student who went to talk to the teacher also read the book would be 1 in 26, because only 1 student from the 26 possible students who went to talk to the teacher read the book.

The correct answer is D.

Question 17

The price of the new bridge is supposed to be 15% less than the price paid to build the bridge 15 years ago. This means you need to multiply the price of the old bridge by 0.85.

$$125,000(0.85) = 106,250$$

The correct answer is A.

Question 18

The mean is the mathematical average of a group of numbers. The median is the middle number of a group of numbers, once all of the numbers are placed in numerical order. If the median is higher than the mean it means, there are more companies producing at or above 2 million washers per day. The mean being so low means that there are very few companies who produce well below 1.5 million washers in order to bring the average down.

The correct answer is B.

Question 19

To find the average number of electronic devices owned by the 250 students surveyed at college X, add up the total number of electronic devices owned by the 250 surveyed students at college X and divide that number by 250.

$$\frac{(1)25 + (2)40 + (3)60 + (4)85 + (5)40}{250} = \frac{825}{250} = 3.3$$

Round the 3.3 to the nearest whole number, which is 3.

The correct answer is B.

Question 20

The entire population at college X is 5,000/250 = 20 times larger than the surveyed population. At college X it would be expected that (20)85 = 1,700 students own 4 electronic devices.

The entire population at college Y is 6,000/250 = 24 times larger than the surveyed population. At college Y is would be expected that (24)70 = 1,680 students own 4 electronic devices.

The sum of the students expected to own 4 electronic devices at college X and college Y is 1,700 + 1,680 = 3,380.

The correct answer is C.

Question 21

The actual time, a needed to study for the test will be 5 days less or 5 days more than the estimate, t, of 120 days. Which means a could be as low as 115 days and as high as 125 days.

For the minimum value of $a = 115$, then $a - t = 115 - 120 = -5$. This means that $a - t$ must be greater than or equal to -5.

For the maximum value of $a = 125$, then $a - t = 125 - 120 = 5$. This means that $a - t$ must also be less than or equal to 5.

The only solution that meets these requirements is A.

The correct answer is A.

Question 22

This question is asking you to solve the gravitational equation in terms of the distance between the two bodies squared, which is r^2. To do this first multiply the entire equation by r^2 so that it is out of the denominator on the right side of the equation.

$$r^2 \left(F = \frac{GmM}{r^2} \right)$$
$$Fr^2 = GmM$$

Now all that is left to do is to divide the entire equation by F, in other words multiply the entire equation by $\frac{1}{F}$.

$$\frac{1}{F}(Fr^2 = GmM)$$
$$\frac{F}{F}r^2 = \frac{GmM}{F}$$
$$r^2 = \frac{GmM}{F}$$

So the above shows the equation solved in terms of r^2.

The correct answer is D.

Question 23

This problem only requires you to plug in the values for the different variables and solve for F. With the variables as follows:

- $G = 6.674 \times 10^{-11}$

- $m = 6 \times 10^7$

- $M = 5 \times 10^5$

- $r = 100$

Plugging these values into the equation results in the following:

$$F = \frac{GmM}{r^2} = \frac{6.674 \times 10^{-11}(6 \times 10^7)(5 \times 10^5)}{100^2} = 0.2$$
$$F = 0.2$$

The correct answer is B.

Question 24

The equation of a circle is:

$$(x-a)^2 + (y-b)^2 = r^2$$

Where:

- a is the x-coordinate of the center of the circle.

- b is the y-coordinate of the center of the circle.

- r is the radius of the circle.

For this question a is -3, $b = -2$ and the radius is 5. Remember that the radius is half of the diameter. Plug those values into the equation for a circle above.

The correct answer is D.

Question 25

One of the points is given with $r = 1$ as (1,0). The other point can be solved by performing some algebra on q. From the equation $r - q = 3$, start by subtracting r from both sides.

$$r - r - q = 3 - r$$
$$-q = 3 - r$$

Now, because the q is negative, multiply the entire equation by negative 1. In practice, you can just change the sign in front of every term of the equation.

$$-1(-q = 3 - r)$$
$$q = r - 3$$

Plugging in the value of $r = 1$, you will find that $q = -2$.

Now you have two points: (1,0) and (0,-2). The slope of a line is the ratio of the change in the y-direction to the change in the x-direction. Mathematically that looks as follows:

$$\frac{0 - (-2)}{1 - 0} = \frac{2}{1} = 2$$

The correct answer is C.

Question 26

The function values referred to in this question are the y-coordinates. The y-coordinate for $g(-\frac{5}{2})$ is found by going to -2.5 on the x-axis and going vertically up until you land on the curve, once on the curve look horizontally to the y-axis to see what the y-value for the curve is as $x = -2.5$. For $x = -2.5$ the y-value is about 2. This value is not less than -2, so I is not a possible answer to this question.

At $g(0)$ the y-value is about -2.2, which is less than -2. This means II is an answer to the question.

At $g(3)$ the y-value is -3 which is also less than -2. Which means III is also an answer to this question.

The correct answer is D.

Question 27

The temperature gradient is defined as the ratio in change in temperature to the change in physical location. Mathematically that looks as follows:

$$\frac{T_2 - T_1}{x_2 - x_1}$$

Where T_2 is the temperature at x_2 and T_1 is the temperature at x_1. For this question the changes in distance are conveniently 1. So that all you need to look at is the change in temperature between any two points. Between $x = 1$ and $x = 2$ the temperature change is larger for material B than for material A. This means that the temperature gradient is larger for material B than for material A over this section.

Note that anytime you measure the temperature change over the *same* distance for both materials, then any comparisons in the temperature difference will lead you to the same conclusions for the temperature gradient.

The conclusion we came to for the temperature gradient over the region from $x = 1$ to $x = 2$, already eliminates choices A and D from being possible answers.

Now look at the region from $x = 4$ and $x = 4$, over this region the temperature change is larger for material A than for material B. This eliminates answer B as being correct.

The correct answer is C.

Question 28

The equation for the area of an isosceles triangle is:

$$A = \frac{1}{2}bh$$

Where:

- A is the area of the triangle.

- b is the length of the base of the triangle.

- h is the height of a line which bisects the triangle.

The base of the triangle is given as $\sqrt{18}$. The height needs to be calculated. The point (1,4) will help you calculate the side of the triangle which spans from the origin to the point (1,4), this point is also a vertex connected to the base. So, in order to calculate this side length (from the origin to (1,4)), you need to realize that this is the hypotenuse of a triangle with side lengths of 4 and 1. The side length is calculated using the equation $a^2 + b^2 = c^2$ solved exclusively for c as $c = \sqrt{a^2 + b^2}$. In this case $a = 4$ and $b = 1$.

$$c = \sqrt{4^2 + 1^2} = \sqrt{16 + 1} = \sqrt{17}$$

Leave c in this form instead of writing down 4.123105626. It's much easier to hold the accuracy by keeping a number as a root. Now this length is the length of the hypotenuse of a triangle whose side lengths are half of the base and the height of the bisection line of the triangle, which is the height you are looking for. Again, you need to use the equation $c^2 = a^2 - b^2$, this time however, you need to solve for either a or b. That is done as follows, try to follow along:

$$a^2 + b^2 = c^2$$
$$b^2 = c^2 - a^2$$
$$b = \sqrt{c^2 - a^2}$$

Now you need to know that c is $\sqrt{17}$ as calculated already. In this equation a will equal half of the base length or $\frac{\sqrt{18}}{2}$. With these values calculate the height as follows:

$$b = \sqrt{\sqrt{17}^2 - \left(\frac{\sqrt{18}}{2}\right)^2}$$
$$b = \sqrt{17 - \frac{18}{4}}$$
$$b = \sqrt{12.5}$$

Now you have all the variables you need in order to calculate the area of the triangle. Remember that this question is asking for the area of the entire shaded region, which is twice the area of a single triangle. This means you need to multiply your area equation by 2. The values of the variables in the area equation are as follows:

- $b = \sqrt{18}$

- $h = 12.5$

The area will be calculated as follows:

$$A = (2)\frac{1}{2}\sqrt{18}(12.5) = 53$$

The correct answer is A.

Question 29

The number of solutions to a system of equations is the same as the number of times the curves from each equation intersect with one another. In this case you are dealing with a straight horizontal line with $y = -1$ and a circle with an unknown center point, but with a radius of 3. These two equations will have only one solution where the circle sits just on top of the horizontal line. The x-coordinate can be anything for the circle; however, the y-coordinate needs to be 3 greater than -1, that would be 2.

The correct answer is C.

Question 30

The total volume of this stack of CD's is the volume of the cylinder with the radius of b minus the volume of the cylinder with the radius of a.

The height of the entire stack of CD's is $780(0.0625) = 48.75$.

The volume equation for a cylinder with the subtraction of the inner cylinder looks as follows. Also notice how much can be factored out:

$$V = \pi b^2 h - \pi a^2 h = \pi h(b^2 - a^2)$$

Now plugging in the values for h, a and b the calculation looks as follows:

$$V = \pi 48.75(2^2 - 0.25^2) = 603$$

The correct answer is B.

Question 31

One kilometer is 1,000 meters. If Timothy swims 100 meters per minute and will swim nonstop for 1,000 meters, it will take him 10 minutes.

$$\frac{1 \text{ minute}}{100 \text{ meters}} 1,000 \text{ meters} = 10 \text{ minutes}$$

The correct answer is 10.

Question 32

You are looking for how many feet added to 8 inches makes 188 inches. With 12 inches per foot, divide 188 by 12 to get 15.66. The number of feet is going to be whole number, so multiply the 15 by 12 to see if everything works out.

$$(15)12 = 180$$

180 inches with another 8 inches will be 188 inches.

The correct answer is 15.

Question 33

The question is giving you a value for x, $x = 2$, and a value for $g(x = 2) = -12$. All you need to do is solve the function for c and plug in the values for x and $g(x = 2)$. You can perform this algebra in one giant chunk to save time by subtracting $2x^3$, $3x$ and adding 14 to both sides of the equation.

$$g(x) - 2x^3 - 3x + 14 = -2x^3 + 2x^3 + 3x - 3x - 14 + 14 - cx^2$$
$$g(x) - 2x^3 - 3x + 14 = -cx^2$$

Now divide the entire equation by $-x^2$, in other words multiply the entire equation by $-\dfrac{1}{x^2}$

$$-\frac{1}{x^2}(g(x) - 2x^3 - 3x + 14 = -cx^2)$$
$$-\frac{g(x) - 2x^3 - 3x + 14}{x^2} = \frac{-x^2}{-x^2}c$$
$$-\frac{g(x) - 2x^3 - 3x + 14}{x^2} = c$$

Now plug in the values for x and $g(x)$ and solve for c.

$$c = -\frac{-12 - 2(2^3) - 3(2) + 14}{2^2} = -\frac{-12 - 16 - 6 + 14}{2^2} = \frac{-20}{4} = -5$$
$$c = -5$$

The correct answer is -5.

Question 34

This question wants you to write down and solve a system of equations. Chuck, who we will call C, and Sarah, who we will call S, ate a combined 370 cheeseburgers. Mathematically that looks as follows:

$$C + S = 370$$

Sarah ate 70 more cheeseburgers than Chuck, mathematically that looks as follows:

$$S + 70 = C$$

You are looking for the number of cheeseburgers Sarah ate, which means when you write your system of equations you will need to solve for C immediately and place it into the other equation.
The second equation, $S + 70 = C$, is already solved for C. Plug that into the top equation and solve for S.

$$C + S = 370$$
$$S + 70 + S = 370$$

Combine the S variables and subtract 70 from both sides of the equation.

$$2S + 70 - 70 = 370 - 70$$
$$2S = 300$$

All that is left to do is to divide the entire equation by 2. In other words multiply the entire equation by $\frac{1}{2}$.

$$\frac{1}{2}(2S = 300)$$
$$\frac{2}{2}S = \frac{300}{2}$$
$$S = 150$$

The correct answer is 150.

Question 35

Every month, m, Stuart added a certain amount of coins to his collection. The number of coins he added every month must be the number which is multiplied by the number of months. If c represents the total number of coins collected, then 23 must be the number of coins he started out with.

The correct answer is 23.

Question 36

The angle of a straight line is 180°. The angle opposite to 165° is 180° - 165° = 15°. The angle opposite to 115° is 180°- 115° = 65°. The sum of all the angles of a triangle is 180°. So the value of A is 180° - 15° - 65° = 100°.

The correct answer is 100.

Question 37

In order to solve for x_n you need to plug the value of $x_{n-1} = 2$ into $f(x_{n-1})$ and $f'(x_{n-1})$. That will look as follows:

$$f(x_{n-1} = 2) = 3(2^3) + 2(2^2) + 1 = 24 + 8 + 1 = 33$$
$$f'(x_{n-1}) = 9(2^2) + 4(2) = 36 + 8 = 44$$

Plug these values into the equation x_n and solve.

$$x_n = x_{n-1} - \frac{f(x_{n-1})}{f'(x_{n-1})} = 2 - \frac{33}{44} = 1.25$$
$$x_n = 1.25$$

the correct answer is 1.25.

Question 38

Solve the following equation for $f'(x_{n-1})$:

$$x_n = x_{n-1} - \frac{f(x_{n-1})}{f'(x_{n-1})}$$

Start by subtracting x_{n-1} from both sides.

$$x_n - x_{n-1} = x_{n-1} - x_{n-1} - \frac{f(x_{n-1})}{f'(x_{n-1})}$$
$$x_n - x_{n-1} = -\frac{f(x_{n-1})}{f'(x_{n-1})}$$

Now multiply the entire equation by $f'(x_{n-1})$.

$$f'(x_{n-1})\left(x_n - x_{n-1} = -\frac{f(x_{n-1})}{f'(x_{n-1})} \right)$$
$$f'(x_{n-1})(x_n - x_{n-1}) = -f(x_{n-1})$$

Now divide the entire equation by $(x_n - x_{n-1})$, in other words multiply the entire equation by $\frac{1}{(x_n - x_{n-1})}$.

$$\frac{1}{(x_n - x_{n-1})}(f'(x_{n-1})(x_n - x_{n-1}) = -f(x_{n-1}))$$
$$f'(x_{n-1}) = -\frac{f(x_{n-1})}{x_n - x_{n-1}}$$

With $x_{n-1} = 3$, $x_n = 2$ and $f(x_{n-1}) = 5x_{n-1} + 4 = (5)3 + 4 = 19$, you can solve the above equation for $f'(x_{n-1})$.

$$f'(x_{n-1}) = -\frac{f(x_{n-1})}{x_n - x_{n-1}} = -\frac{19}{2-3} = -\frac{19}{-1} = 19$$
$$f'(x_{n-1}) = 19$$

The correct answer is 19.

Chapter 8

Practice Test 2

Practice Math Test - No Calculators Allowed

Directions

You will have 25 minutes to complete part 3 of the math practice test. There are 20 questions in this portion of the test. Questions 1-15 are multiple choice and questions 16-20 are bubble-in questions where no choices are provided for you. You may use a piece of scratch paper for notes and calculations. Write your answers down on a separate piece of paper, the solutions and explanations can be found immediately after the test. You are to continue when you see the word CONTINUE in the bottom right corner of the page. Stop taking the test when the word STOP appears in the bottom center of the page. If you finish before the time expires you may check your answers for part 3 of the test only, do not move ahead within the 25 minutes provided for the portion of the test.

Notes

- No calculator usage allowed

- All variables and expressions belong to the set of real numbers unless otherwise written.

- Not all figures are drawn to scale. If the scaling is important it will be noted.

- If a drawing is in three-dimensions it will be noted, otherwise all drawings are in two-dimensions.

CONTINUE

Reference Equations

- The area of a circle is $A = \pi r^2$

- The circumference of a circle is $C = 2\pi r$

  r is the radius of the circle

- The area of a rectangle is $A = lw$

  l is the length of the rectangle

  w is the width of the rectangle

- The area of a triangle is $A = \dfrac{1}{2}bh$

  b is the base of the triangle

  h is the height of the triangle

- The sides of a right triangle have the following relationship: $a^2 + b^2 = c^2$

  c is the length of the hypotenuse

  a and b are the side lengths for the sides whose ends meet at the right angle

- A right triangle with angles of $30°$, $60°$ and $90°$ angles has side lengths of x and $x\sqrt{3}$ and a hypotenuse length of $2x$.

- A right triangle with angles of $45°$, $45°$ and $90°$ angles has side lengths of x and x and a hypotenuse length of $2x\sqrt{2}$.

- The volume of a rectangular prism is $V = lwh$

  h is the height of the prism

- The volume of a cylinder is $V = \pi r^2 h$

- The volume of a sphere is $V = \dfrac{4}{3}\pi r^3$

- The volume of a cone is $V = \dfrac{1}{3}\pi r^2 h$

- The volume of a pyramid is $V = \dfrac{1}{3}lwh$

- There are $360°$ in the arc of a circle

- There are 2π radians in the arc of a circle

- The sum of all the angles in a triangle equals $180°$

CONTINUE

1.

If $\dfrac{K+2}{y} - 4 = 3$ and $K = 19$, what is the value of y?

A) 25

B) 3

C) 4

D) 19

2.

Which of the following is the sum of $(6+2j)-(4-1j)$? (Note: $j = \sqrt{-1}$)

A) $2 + j$

B) $8 - 5j$

C) $10 + 3j$

D) $2 + 3j$

3.

Suzy came down with a cold. She was sick for the next d days. During this time, she sneezed an average of 6 times per hour and coughed an average of 14 times per hour. If h represents the number of hours Suzy was awake every day while she was sick, and Suzy did not sneeze or cough during her sleep, what expression represents the total number of times Suzy sneezed and coughed while she had a cold?

A) 6d + 14h

B) 6h + 14d

C) d(6h + 14h)

D) h(6d + 14h)

4.

Charles needs to pick up his daughter from school 5 days a week. Every minute, m, he walks 150 meters. During his walk the distance remaining, d, to the school can be described with the relationship $d = 1950 - 150m$. What is the best interpretation of the value 1950?

A) The time it takes Charles to walk from his house to the school

B) The distance Charles walks every minute

C) The distance Charles walks in five minutes

D) The total distance from Charles' house to the school

CONTINUE

5.

$$5s^4 - 3t + 6s^3t^2 - (-7t + s^4 - s^3t^2)$$

Which of the following is equivalent to the expression above?

A) $4s^4 + 4t + 7s^3t^2$

B) $4s^4 + 4t + 7s^3 + 2t^2$

C) $6s^4 - 10t + 5s^3t^2$

D) $6s^4 - 10t + 5s^3$

6.

$$P = 72 - 0.2x$$

Irrigation system designers want to estimate the pressure of a system, P, at a certain distant, x, downstream of the filter. Based on the model what is the expected loss in pressure every foot?

A) 72

B) 0.2

C) 14.4

D) 71.8

7.

$$R = \frac{\left(1 - \dfrac{2 - \pi}{3}\right)^t}{14.7 \times 10^{-14} - \dfrac{\pi^2}{64}} Q$$

An engineering student developed the equation for R in terms of t and Q. How could he rewrite the equation for Q in terms of t and R?

A) $Q = \dfrac{\left(14.7 \times 10^{-14} + \dfrac{\pi^2}{64}\right)}{\left(1 - \dfrac{2 - \pi}{3}\right)^t} R$

B) $Q = \dfrac{\left(14.7 \times 10^{-14} - \dfrac{\pi^2}{64}\right)}{\left(1 - \dfrac{2 - \pi}{3}\right)^t} R$

C) $Q = \dfrac{\left(14.7 \times 10^{-14} - \dfrac{\pi^2}{64}\right)}{\left(1 - \dfrac{2 - \pi}{3}\right)^R} t$

D) $\left(14.7 \times 10^{-14} - \dfrac{\pi^2}{64}\right) R - \left(1 - \dfrac{2 - \pi}{3}\right)^t$

CONTINUE

8.

If $\dfrac{1}{xy} = 5$, what is $10xy$?

A) 1

B) 2

C) 3

D) 4

9.

$$\frac{1}{2}x + \frac{1}{3}y = -1$$
$$3x - 2y = 18$$

Which of the following is a solution (x, y) to the system of equations above?

A) $(2,6)$

B) $(2,-6)$

C) $(-2,6)$

D) $(-2,-6)$

10.

$$s(t) = t^2 + 3$$

The function above defines s in terms of t. If $s(2) = 11$, what is the value of $s(-2)$?

A) 7

B) 11

C) -7

D) -11

11.

$$H = 30 + 13z$$
$$B = 150 + z$$

The equations above show the number of hawks, H, and bluebirds, B, sitting on a telephone wire at a distance x away from the first telephone pole. How long would the telephone wire need to be so that the same number of hawks and bluebirds would be perched upon the wire?

A) 8

B) 9

C) 10

D) 11

12.

A line on the xy-plane passes through the point $(3,1)$ and passes through the origin. Which of the following is the best choice for the slope of the line?

A) $\frac{1}{3}$

B) -3

C) -1

D) 1

CONTINUE

13.

if t > 5, which of the following expressions is equivalent to the expression below?

$$\frac{2}{\dfrac{1}{t-4}+\dfrac{3}{t+1}}$$

A) $2t^2 - 6t - 8$

B) $4t - 11$

C) $\dfrac{4t - 11}{2t^2 - 6t - 8}$

D) $\dfrac{2t^2 - 6t - 8}{4t - 11}$

14.

If $2a + 3b = 13$, what is the value of $3^{2a}9^b$?

A) 3^{13}

B) 9^2

C) 13^3

D) There is not enough information to determine the solution to this problem.

15.

$$(2t + 3)(t - 4) = at^2 + bt + c$$

If the above equation is true for all values of t. What are the values of the constants a, b and c?

A) -2, 5 and 12

B) 2, 11 and 12

C) 2, -5 and -12

D) -2, 11 and -12

CONTINUE

Directions

The final questions in part 3 of the test are to be bubbled in. On the actual test there will be pictures to show you how to properly bubble in the answers on your answer sheet. For this book, you only need to write down your answers on a separate sheet of paper. You will be able to bubble in fractions and decimals on the actual test, so don't worry if answers appear in this form on the practice test.

CONTINUE

16.

If $\sqrt[3]{t^2} = \dfrac{3}{3}$ and $t > 0$ what is the value of t?

18.

$$7q + 3r = -5$$
$$-3q + 5r = 21$$

According to the system of equations above, what is the value of r?

17.

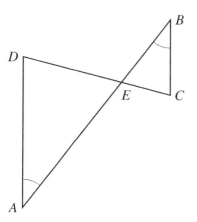

In the graphic above the lengths $\overline{AE}$, $\overline{EB}$ and $\overline{DE}$ are 120, 40 and 100 respectively. The angles $\angle EBC$ and $\angle EAD$ are the same. What is the length of $\overline{EC}$?

19.

In a right triangle the measure of an angle is θ, where $\cos(\theta) = \dfrac{2}{3}$. What is the value of $\sin(\theta + \dfrac{\pi}{2})$?

20.

If $z = 2^x$, where $x > 0$. What is the value of $\sqrt[x]{z}$?

STOP

If you finished early, check only your work for this section.
Do not look at any other section.

Directions

You will have 55 minutes to complete part 4 of the math practice test. There are 38 question on this portion of the test. Questions 1-30 are multiple choice and questions 31-38 are bubble-in questions where no choices are provided for you. You may use a piece of scratch paper for notes and calculations. Write your answers down on a separate piece of paper, the solutions and explanations can be found immediately after the test. You are to continue when you see the word CONTINUE in the bottom right corner of the page. Stop taking the test when the word STOP appears in the bottom center of the page. If you finish before the time expires you may check your answers for part 4 of the test only, do not return to part 3 within the 55 minutes provided for the portion of the test.

Notes

- Calculator usage is allowed

- All variables and expressions belong to the set of real numbers unless otherwise written.

- Not all figures are drawn to scale. If the scaling is important it will be noted.

- If a drawing is in three-dimensions it will be noted, otherwise all drawings are in two-dimensions.

CONTINUE

Reference Equations

- The area of a circle is $A = \pi r^2$

- The circumference of a circle is $C = 2\pi r$

    r is the radius of the circle

- The area of a rectangle is $A = lw$

    l is the length of the rectangle

    w is the width of the rectangle

- The area of a triangle is $A = \dfrac{1}{2}bh$

    b is the base of the triangle

    h is the height of the triangle

- The sides of a right triangle have the following relationship: $a^2 + b^2 = c^2$

    c is the length of the hypotenuse

    a and b are the side lengths for the sides whose ends meet at the right angle

- A right triangle with angles of $30°$, $60°$ and $90°$ angles has side lengths of x and $x\sqrt{3}$ and a hypotenuse length of $2x$.

- A right triangle with angles of $45°$, $45°$ and $90°$ angles has side lengths of x and x and a hypotenuse length of $2x\sqrt{2}$.

- The volume of a rectangular prism is $V = lwh$

    h is the height of the prism

- The volume of a cylinder is $V = \pi r^2 h$

- The volume of a sphere is $V = \dfrac{4}{3}\pi r^3$

- The volume of a cone is $V = \dfrac{1}{3}\pi r^2 h$

- The volume of a pyramid is $V = \dfrac{1}{3}lwh$

- There are $360°$ in the arc of a circle

- There are 2π radians in the arc of a circle

- The sum of all the angles in a triangle equals $180°$

CONTINUE

1.

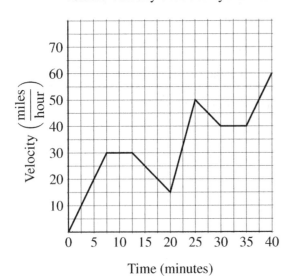

Vehicle Velocity over Fourty Minutes

From the graph above, which of the following time periods describes a time when the vehicle sped up, then drove at a constant speed and then slowed down?

A) 0 minutes to 20 minutes

B) 7.5 minutes to 25 minutes

C) 12.5 minutes to 25 minutes

D) 25 minutes to 40 minutes

2.

For $y = \dfrac{x}{C}$, C is a constant and when $x = 4$ then $y = 16$. What is the value of y when $x = 3$?

A) 6

B) 9

C) 12

D) 15

3.

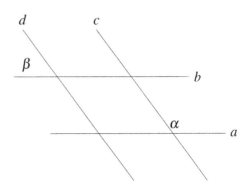

In the graphic above $a \parallel b$ and $c \parallel d$. If the measure of $\angle\alpha$ is 52.5°, what is the measure of $\angle\beta$?

A) 37.5°

B) 127.5°

C) 105°

D) 52.5°

4.

$$3y - 17 = 4$$

From the equation above, what is the value of $9y$?

A) 17

B) 7

C) 63

D) 21

CONTINUE

5.

Which of the following is an example of no correlation?

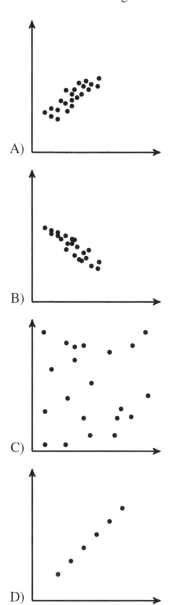

6.

0.001 Newtons = 1 Millinewton

1000 Newtons = 1 Kilonewton

During a metal rod stress test at the local university a force of 500 Millinewtons was measured inside the rod. What is this measurement in Kilonewtons?

A) 0.5

B) 0.05

C) 0.005

D) 0.0005

CONTINUE

7.

Leaves on Front Yards of Four Houses

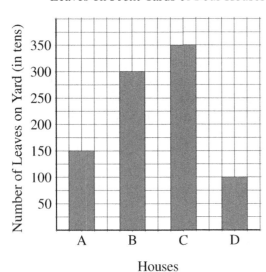

A very curious kid went around and counted the number of leaves on the lawns from four of his neighbors. The data is collected in the bar graph above. What is the average numbers of leaves found on his neighbors' lawns?

A) 2250

B) 225

C) 1500

D) 150

8.

For which value of q is $|q-5|-5$ equal to -2?

A) -8

B) 5

C) 2

D) There is no such value of q.

CONTINUE

The following information is for questions 9 and 10

$$\frac{v^2\rho}{2} = \frac{w^2\rho}{2} + P$$

The above equation is a relationship between properties of a moving fluid at two different points along its path. In this equation ρ is the density of the fluid, v is the velocity of the fluid at the first point, w is the velocity of the fluid at the second point and P is the pressure of the fluid at the first point.

9.

Which of the following equations expresses the pressure in terms of the fluid velocities and density?

A) $P = \dfrac{2}{\rho}(w^2 - v^2)$

B) $P = \dfrac{\rho}{2}(v^2 - w^2)$

C) $P = \dfrac{2}{\rho}(v^2 - w^2)$

D) $P = \dfrac{\rho}{2}(w^2 - v^2)$

10.

The equation above written for v^2 is

$$v^2 = w^2 + \frac{2P}{\rho}$$

If $w^2 = 50$ and $\dfrac{P}{\rho} = 25$, what is the value of v?

A) 10

B) 75

C) 100

D) 150

11.

$$\frac{y}{2} + 14 \le 2y + 8$$

Which of the following is a solution of the above inequality?

A) 7

B) 6

C) 5

D) 4

12.

Cars Sold by Different Salespeople

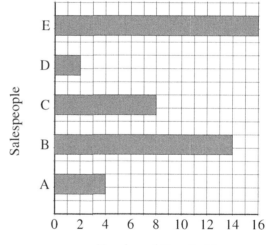

What is the median number of cars sold by these salespeople?

A) 8.8

B) 8

C) 2

D) 16

CONTINUE

13.

	Bonds	Stocks	Real Estate	Total
Doctor	13	52	21	86
Lawyer	20	78	9	107
Total	33	130	30	193

Groups of doctors and lawyers were asked how they invest the money from their exorbitant incomes. Their responses were collected in the table above. Which group makes up 11 % of all the doctors and lawyers from this survey?

A) Doctors who buy bonds.

B) Lawyers who buy stocks.

C) Doctors who buy real estate.

D) Lawyers who buy bonds.

14.

The following table presents the number of bacteria in the petri dishes of every student from a tenth-grade biology class.

14	67	03	22	10	34	77	40	51
22	39	72	11	06	63	41	81	58

Which of the following is the range of bacteria found in the petri dishes of the tenth-grade students?

A) 81

B) 78

C) 3

D) 42

15.

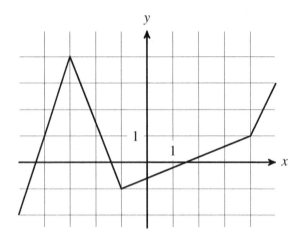

Shown in the graphic above is the entire curve for the function $f(x)$. What is the minimum function value?

A) -5

B) 4

C) -1

D) -2

CONTINUE

The following information is for questions 16 and 17.

Plant Growth Relationship to Hours of Sunlight

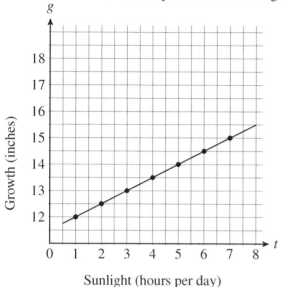

The graph above shows the relationship between the growth of plants, g, and the hours of sunlight they receive every day, t.

16.

How many hours of sunlight per day would your plant need in order for it to grow to be 14.5 inches?

A) 4

B) 5

C) 6

D) 7

17.

Which of the following best represents the function $g(t)$ as shown in the graph?

A) $g = 12 + \dfrac{1}{2}t$

B) $g = 11.5 + \dfrac{1}{2}t$

C) $g = 11.5 + 2t$

D) $g = 12 + 2t$

CONTINUE

18.

$$y \geq 3x + 8$$
$$y - 3 \geq 2x + 4$$

For the range of $-5 \leq x \leq 5$, which of the following values for y is NOT a solution to the above system of inequalities?

A) -3

B) 18

C) 23

D) -8

19.

A steel manufacturer sells black steel straight pipe for $2.50 per foot and stainless-steel straight pipe for $7.50 per foot. The steel manufacturer made $945.00 from selling a total of 226 feet of black, b, and stainless, s, steel pipe. Which of the following system of equations can be used to solve for both lengths of pipe sold?

A) $\quad \dfrac{b}{2.5} + \dfrac{s}{7.5} = 945$
$\qquad\qquad b + s = 226$

B) $\quad 2.5b + 7.5s = 945$
$\qquad\qquad b + s = 226$

C) $\quad \dfrac{b}{2.5} + \dfrac{s}{7.5} = 226$
$\qquad\qquad b + s = 945$

D) $\quad 2.5b + 7.5s = 226$
$\qquad\qquad b + s = 945$

CONTINUE

20.

Julie's mom bought her a pack of 60 dominoes. Julie took her dominoes to the park and received some dominoes from a stranger, which increased the amount of dominoes she had by 20%. The next day she took all her dominoes to school and lost 33% of them. How many dominoes did Julie have afterwards?

A) 24

B) 48

C) 72

D) 40

21.

Bowls of cereal eaten by college students in the morning

	1	2 to 3	4 or more	Total
Female Students	23	36	11	70
Male Students	5	46	19	70
Total	28	82	30	140

If a male student, who took part in this survey, was selected at random, what would be the probability that they eat 2 or less bowls of cereal in the morning?

A) $\dfrac{51}{110}$

B) $\dfrac{51}{140}$

C) $\dfrac{59}{110}$

D) $\dfrac{51}{70}$

CONTINUE

The emergency department transfers patients to different departments of the hospital. The following table is a collection of data showing where patients were transferred to, throughout the hospital, during the busiest months of the year.

Department-Month	January	February	March	April	May	July	September	November	December
Cardiology	5	14	25	5	5	15	7	19	16
Pediatrics	14	6	8	20	9	7	14	9	17
Neurology	14	25	8	20	9	14	19	11	15
Oncology	17	13	10	7	14	8	12	18	19
Obstetrics	9	17	19	23	15	24	9	21	15

22.

Which of the following best approximates the average rate of change in patients sent to the oncology department between the months from January to April?

A) 10

B) 4

C) 3

D) 3.3

23.

Of all the departments which department received the most patients in April?

A) Pediatrics

B) Neurology

C) Oncology

D) Obstetrics

CONTINUE

24.

For the equation of a circle $(x-3)^2 + (y+2) = 9$, which of the following are the center point, C, and radius, r, of this circle?

A) $C = (-3, 2), r = 9$

B) $C = (3, -2), r = 3$

C) $C = (-3, 2), r = 3$

D) $C = (3, -2), r = 9$

25.

$$F = -64x + 320$$

The above equation represents the forces acting on a crane. The force, F, is required to keep the crane from tipping over. The value 64 is the weight on the crane, a distance, x, from the crane base. The value 320 is the weight of the crane. How large does x need to be for F to disappear and the crane to tip?

A) 5

B) 256

C) 20480

D) 10

26.

Both Richard and Kathrine are training for a marathon. On average Kathrine finishes a mile 10% faster than Richard. If Richard runs his mile in 320 seconds, which of the following expressions describes how long it takes Kathrine to run her mile?

A) $320(0.10)$

B) $320(1.10)$

C) $\dfrac{320}{.90}$

D) $320(0.90)$

27.

A curious student threw a bucket of marbles over the entire gymnasium floor. The floor was divided into 16 evenly spaced squared and one student was assigned to each square to count the number of marbles there. The data collected is summarized in the table below:

Square	# of Marbles	Square	# of Marbles
1	21	9	26
2	26	10	21
3	22	11	22
4	28	12	26
5	24	13	28
6	26	14	21
7	28	15	28
8	29	16	23

Approximately, what were the total number of marbles in the bucket?

A) 40

B) 400

C) 20

D) 200

CONTINUE

28.

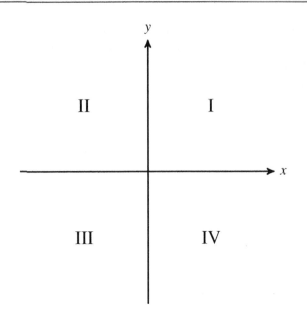

The function $y + 1.75 = -3(x - 1.5)^2$ has its maximum value in which of the quadrants shown above?

A) I

B) II

C) III

D) IV

29.

For the polynomial $y = (x + 4)^2 - 2$, what is the sum of the vertex coordinates multiplied by 2?

A) -12

B) -6

C) 12

D) 6

30.

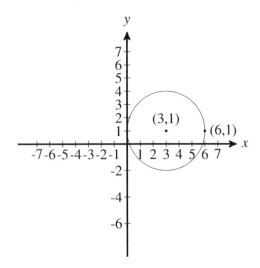

What is the equation for the circle depicted on the graph above?

A) $(x - 3)^2 + (y - 1)^2 = 9$

B) $\dfrac{x^2}{3} + y^2 = 9$

C) $\dfrac{x^2}{3} + y^2 = 3$

D) $(x - 3)^2 + (y - 1)^2 = 3$

CONTINUE

Directions

The final questions in part 4 of the test are to be bubbled in. On the actual test there will be pictures to show you how to properly bubble in the answers on your answer sheet. For this book, you only need to write down your answers on a separate sheet of paper. You will be able to bubble in fractions and decimals on the actual test, so don't worry if answers appear in this form on the practice test.

CONTINUE

31.

On his slowest day Geoff can put together 42 pizza boxes per minute. On his fastest day he can put together 78 pizza boxes per minute. If a shipment of 3,000 pizza boxes came in, what is a possible length of time it could take Geoff to put together the boxes, in hours?

32.

A string will snap when it is put under 200 pounds of tension. The string weighs 0.25 pounds per inch. How long does the string need to be in inches before it snaps under its own weight?

33.

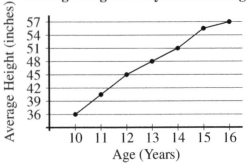

Average Height of Boys Based on Age

What multiple of a 10-year-old average boy's height is the average height of a 14-year-old boy?

34.

A truck can fit 14 refrigerators per load, and it takes the truck exactly 45 minutes to go drop off the load, return and pack up another 14 refrigerators. If the truck delivered refrigerators for 48 hours straight, how many could it deliver?

CONTINUE

35.

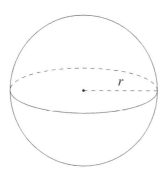

The volume of the sphere depicted above is 288π. What is the diameter of the sphere?

36.

$$H(t) = \frac{1}{t^2 + 5t + 6}$$

What is one of the possible values of t which renders the above equation undefined?

The following information is for questions 37 and 38

A researcher is observing a bacteria farm which started out with 365 bacteria in a petri dish. To estimate the number of bacteria in the dish after d days, the researcher uses the expression $365(b)^d$. The researcher notices that the bacteria grow at a rate of 12% per day.

37.

What is the value of b?

38.

Another researcher started a bacteria farm with 400 bacteria. This farm also grew at a rate of 12% per day. After 2 weeks how many more bacteria will be present in this farm, compared to the farm which started with 365 bacteria? (Round your answer to the nearest whole number)

STOP

If you finished early, check only your work for this section.
Do not look at any other section.

8.1 Practice Test 2 Solutions

Part 3: Multiple Choice

1. B
2. D
3. A
4. D
5. A
6. B
7. B
8. B
9. B
10. B
11. C
12. A
13. D
14. A
15. C

Part 3: Bubble In

16. 1
17. 33.3
18. 3
19. $\frac{2}{3}$
20. 2

Part 4: Multiple Choice

1. A
2. C
3. B
4. C
5. C
6. D
7. B
8. C
9. B
10. A
11. D
12. B
13. C
14. B
15. D
16. C
17. B
18. D
19. B
20. B
21. D
22. D
23. D
24. B
25. A
26. D
27. B
28. D
29. A
30. A

Part 4: Bubble In

31. 41.7 - 71.4
32. 800
33. 1.4
34. 896
35. 12
36. -2 or -3
37. 1.12
38. 171

8.2 Practice Test 2 Explanations

Question 1

The value K is given as 19, which turns this into a problem of solving for y. the first thing to do to solve for y is add 4 to both sides of the equation.

$$\frac{K+2}{y} - 4 + 4 = 3 + 4$$

$$\frac{K+2}{y} = 7$$

Now multiply the entire equation by y.

$$y\left(\frac{K+2}{y} = 7\right)$$

$$K + 2 = 7y$$

With $K = 19$, the equation turns into:

$$21 = 7y$$

Now divide the entire equation by 7, in other words multiply the entire equation by $\frac{1}{7}$.

$$\frac{1}{7}(21 = 7y)$$

$$\frac{21}{7} = \frac{7}{7}y$$

$$3 = y$$

The correct answer is B.

Question 2

When you add or subtract two complex numbers. The real component is added or subtracted with the real component of the second complex number and the complex component is added or subtracted with the complex component of the second complex number.

$$(6+2j) - (4-j) = 6 + 2j - 4 + j = 6 - 4 + 2j + j = 2 + 3j$$

The correct answer is D.

Question 3

Suzy sneezed an average of 6 times per hour while she was awake with a cold. The total number of times she sneezed during her cold is the average number of times she sneezed per hour multiplied by the number of hours that can be written mathematically as $6h$. The total number of times she coughed during her cold is the average number of times she coughed per hour, 14, multiplied by the number of hours she was sick, that can be written mathematically as $14h$. The total number of times she sneezed and coughed is the sum of these two products, or $6h + 14h$.

The correct answer is A.

Question 4

When Charles just starts his walk m will equal zero. When $m = 0$ the distance, d, remaining is $d = 1{,}950 - 150(0) = 1{,}950$. This means 1,950 is the total distance Charles needs to walk.

The correct answer is D.

Question 5

When you are confronted with a giant jumble of terms, the best thing to do is to group the like terms within the sum. In this case before you do that you should distribute the negative sign.

$$5s^4 - 3t + 6s^3t^2 - (-7t + s^4 - s^3t^2) = 5s^4 - 3t + 6s^3t^2 + 7t - s^4 + s^3t^2$$

Then group the like terms as follows:

$$5s^4 - s^4 - 3t + 7t + 6s^3t^2 + s^3t^2$$

Then combine the like terms as follows:

$$4s^4 + 4t + 7s^3t^2$$

The correct answer is A.

Question 6

The expected loss per foot is something that will change every foot traveled. The number 72 cannot change because it is not multiplied by anything. However, the value 0.2 is subtracted with every foot, x, traveled. This value of 0.2 must be the expected pressure loss every foot.

The correct answer is B.

Question 7

This question looks more complicated than it actually is. All you need to do is take the inverse of the fraction in front of Q and place it in front of R. The inverse of the fraction in front of Q is:

Note that the exponent of negative one means take the inverse of whatever is in parenthesis.

$$\left(\frac{\left(1 - \frac{2 - \pi}{3}\right)^t}{14.7 \times 10^{-14} - \frac{\pi^2}{64}} \right)^{-1} = \frac{\left(14.7 \times 10^{-14} - \frac{\pi^2}{64}\right)}{\left(1 - \frac{2 - \pi}{3}\right)^t}$$

This giant fraction is in front of R in answer B.

The correct answer is B.

Question 8

For this problem you need to solve for xy and then multiply it by 10. In order to solve for xy multiply the entire equation by xy.

$$xy\left(\frac{1}{xy} = 5\right)$$
$$1 = 5xy$$

Now divide the entire equation by 5, in other words multiply the entire equation by $\frac{1}{5}$. This will get xy all by itself.

$$\frac{1}{5}(1 = 5xy)$$
$$\frac{1}{5} = xy$$

This number multiplied by 10 is $\frac{10}{5} = 2$.

The correct answer is B.

Question 9

Before jumping into solving this system of equations, notice that all the solutions are either positive or negative 2 and 6. Plug those values into the bottom equation until you get the correct solution. You will find that the solution set is (2,-6), this trial and error will be way faster than attempting to solve the system of equations algebraically.

The correct answer is B.

Question 10

Notice here that $2^2 = (-2)^2$, which means that $s(2) = s(-2)$.

The correct answer is B.

Question 11

If the number of bluebirds and hawks were the same, then B would equal H. Which means you would set the right side of both equations equal to one another.

$$30 + 13z = 150 + z$$

To solve for z, first move all of the z-terms to the left side, by subtracting z from both sides.

$$30 + 13z - z = 150 + z - z$$
$$30 + 12z = 150$$

Now move the constant term 30 to the right side of the equation by subtracting 30 from both sides.

$$30 - 30 + 12z = 150 - 30$$
$$12z = 120$$

Now divide both sides of the equation by 12, in other words multiply the entire equation by $\frac{1}{12}$.

$$\frac{1}{12}(12z = 120)$$

$$\frac{12}{12}z = \frac{120}{12}$$

$$z = 10$$

The correct answer is C.

Question 12

The line passes through the origin and $(3,1)$. So, starting at $(0,0)$ you rise by 1 to get to the point $(0,1)$, from there you run by 3 to get to the point $(3,1)$. Now if the rise is 1 and the run is 3, then the rise over run is $\frac{1}{3}$. The rise over run is also the slope.

The correct answer is A.

Question 13

Focus on the denominator first. Take that sum of fractions down there and place them under a common denominator. To do that multiply 3 by $t - 4$ and multiply 1 by $t + 1$. The common denominator will become the product of $t - 4$ and $t + 1$. Look at how this unfolds mathematically as follows:

$$\frac{1}{t-4} + \frac{3}{t+1} = \frac{(t+1)}{(t+1)(t-4)} + \frac{3(t-4)}{(t+1)(t-4)}$$

Notice how the $t + 1$ and $t - 4$ terms will cancel and produce the left side of the equation. Now combine those two fractions from the right side and make them into a single fraction under a common denominator.

$$\frac{(t+1)}{(t+1)(t-4)} + \frac{3(t-4)}{(t+1)(t-4)} = \frac{(t+1)+3(t-4)}{(t+1)(t-4)}$$

This will change the overall fraction from the question into the following:

$$\frac{2}{\dfrac{(t+1)+3(t-4)}{(t+1)(t-4)}}$$

This is equivalent to:

$$\frac{2}{1} \div \frac{(t+1)+3(t-4)}{(t+1)(t-4)} = \frac{2}{1}\left(\frac{(t+1)(t-4)}{(t+1)+3(t-4)}\right) = \frac{2(t+1)(t-4)}{(t+1)+3(t-4)}$$

You only need to multiply out either the numerator or the denominator to see which answer is correct. In this case the denominator will be easier to work with. Distribute the 3 and combine the likes terms as follows:

$$t + 1 + 3(t - 4) = t + 1 + 3t - 12 = 4t - 11$$

This matches with the denominator of choice D.

The correct answer is D.

Question 14

The expression $3^{2a}9^b$ can be rearranged as follows (refer to the chapter 2, section 2.5 for reference).

$$3^{2a}9^b = 3^{2a}(3^3)^b = 3^{2a}3^{3b} = 3^{2a+3b} = 3^{13}$$

Try practicing this on your own by referring to rules 3 and 1 from section 2.5.

The correct answer is A.

Question 15

The trick with this question is to multiply out the left side of the equation and then do a coefficient comparison. Multiply out the left side of the equation using the F.I.O.L. method. That will look as follows:

$$(2t+3)(t-4) = 2t^2 + 3t - 8t - 12 = 2t^2 - 5t - 12$$

If you compare the coefficients in front of the t^2, t and constant terms. You will see that $a = 2$, $b = -5$ and $c = -12$.

The correct answer is C.

Question 16

The equivalent form of $\sqrt[3]{t^2}$ is $t^{\frac{2}{3}}$. The value of $\frac{3}{3}$ is 1. The important thing to remember here is that 1 raised to any power is still 1. This is also true for the case of $1^{\frac{2}{3}}$. The value of t is 1.

the correct answer is 1.

Question 17

This is a case of similar triangles. This means the ratios of like side lengths from both triangles are equal to the ratios of other like side lengths. Mathematically the following relationships are true:

$$\frac{\overline{EB}}{\overline{AE}} = \frac{\overline{EC}}{\overline{DE}}$$

Solve this equation for the unknown length $\overline{EC}$, by multiplying the entire equation by $\overline{DE}$.

$$\overline{DE}\left(\frac{\overline{EB}}{\overline{AE}} = \frac{\overline{EC}}{\overline{DE}}\right)$$
$$\overline{DE}\left(\frac{\overline{EB}}{\overline{AE}}\right) = \overline{EC}$$

Taking the values from the problem $\overline{AE} = 120$, $\overline{EB} = 40$ and $\overline{DE} = 100$; plug those into the equation and solve for $\overline{EC}$.

$$\overline{EC} = 100\left(\frac{40}{120}\right) = 100\left(\frac{1}{3}\right) = 33.3$$

The correct answer is 33.3.

Question 18

To solve for r quickly solve one of the equations for q and plug it into the other equation. Solving either equation for q is not going to be pretty, go ahead and solve the bottom equation for q. This decision was made because the q in the bottom equation is multiplied only by -3 whereas the q in the top equation is multiplied by 7. To solve for q in the bottom equation first subtract $5r$ from both sides.

$$-3q + 5r - 5r = 21 - 5r$$
$$-3q = 21 - 5r$$

Now divide the entire equation by -3, in other words multiply the entire equation by $-\frac{1}{3}$.

$$-\frac{1}{3}(-3q = 21 - 5r)$$
$$q = -\frac{21 - 5r}{3}$$

Plug this value for q into the top equation.

$$7\left(-\frac{21 - 5r}{3}\right) + 3r = -5$$

The multiplication between 7 and 21 can be quite cumbersome without a calculator; however, with some rearranging you will see that this multiplication can fall away. First bring the negative sign in front of the 7 and then separate the single fraction in parenthesis into a sum of fractions. Let's look only at the left side of the equation.

$$7\left(-\frac{21 - 5r}{3}\right) + 3r = -7\left(\frac{21}{3} - \frac{5r}{3}\right) + 3r$$

Now 21 divided by 3 is something that is much more manageable, that is equal to 7. Let's see the left side of the equation with this division. Then you can distribute the 7 into the sum in the parenthesis, don't worry about dividing by 3 in the second fraction yet. You will see how to handle that in the next step.

$$-7\left(\frac{21}{3} - \frac{5r}{3}\right) + 3r = -7\left(7 - \frac{5r}{3}\right) + 3r = -49 + \frac{35r}{3} + 3r$$

Now in order to get rid of the three in the denominator, you are going to need to multiply the entire equation by 3.

$$3\left(-49 + \frac{35r}{3} + 3r = -5\right)$$
$$-147 + 35r + 9r = -15$$

If you have made it this far, you almost have it. Next combine the r terms.

$$-147 + 35r + 9r = -15$$
$$-147 + 44r = -15$$

The next step is to add 147 to both sides of the equation.

$$-147 + 147 + 44r = -15 + 147$$
$$44r = 132$$

The final step to solve for r is to divide the entire equation by 44, in other words multiply the entire equation by $\frac{1}{44}$.

$$\frac{1}{44}(44r = 132)$$
$$r = \frac{132}{44}$$
$$r = 3$$

The correct answer is 3.

Question 19

A very important trigonometric relationship to know is:

$$\cos(\theta) = \sin(\theta + \frac{\pi}{2})$$

This relationship means that the sine function is always $90°$ ahead of the cosine function. It also means that the correct answer is given by $\cos(\theta)$. Another important similar trigonometric relationship to remember is:

$$\sin(\theta) = \cos(\theta - \frac{\pi}{2})$$

The correct answer is $\frac{2}{3}$

Question 20

In order to get rid of the exponent x above the 2, you will need to raise the entire equation to $\frac{1}{x}$ power. That will look as follows:

$$(z = 2^x)^{\frac{1}{x}}$$
$$z^{\frac{1}{x}} = (2^x)^{\frac{1}{x}}$$
$$z^{\frac{1}{x}} = 2$$

What are important to remember at this step are the rules of roots, namely, that the following is true:

$$z^{\frac{1}{x}} = \sqrt[x]{z}$$

The correct answer is 2.

Question 1

Because you are looking at a velocity versus time graph, a period where the vehicle sped up, drove at a constant rate and then slowed down will be shown by the curve going diagonally up, followed by a section where the line is horizontal, and finally the line will drop diagonally down. This curve behavior is present during the period from 0 to 20 minutes.

The correct answer is A.

Question 2

When the values of $x = 4$ and $y = 16$ are given, you are able to solve for the value of C. Before plugging in the values, rearrange the equation with the variables to solve for C. Begin by multiplying the entire equation by C.

$$C\left(y = \frac{x}{C}\right)$$
$$Cy = x$$

Next, to get the C all by itself, divide the entire equation by y, in other words multiply the entire equation by $\frac{1}{7}$.

$$\frac{1}{y}(Cy = x)$$
$$C = \frac{x}{y}$$

Now plug in the values for x and y given in the problem, and solve for the value of C.

$$C = \frac{4}{16} = \frac{1}{4}$$

Now because C is in the denominator of the equation $y = \frac{x}{C}$, the following is going to unfold mathematically, which will greatly simplify the problem:

$$y = \frac{x}{C}$$
$$y = \frac{x}{\frac{1}{4}}$$
$$y = 4x$$

For the value of $x = 3$, the equation will look like and can be solved as follows:

$$y = 4(3) = 12$$

The correct answer is C.

Question 3

Because $a \parallel b$ and $c \parallel d$, the angles α and β add up to $180°$. The measure of $\angle\beta$ is $180° - 52.5° = 127.5°$.

The correct answer is B.

Question 4

The first thing to do here is to add 17 to both sides of the equation.

$$3y - 17 + 17 = 4 + 17$$
$$3y = 21$$

Now you can easily get the $3y$ to a $9y$ when you multiply the entire equation by 3.

$$3(3y = 21)$$
$$9y = 63$$

The correct answer is C.

Question 5

No correlation means there is no relationship between the data presented. In three out of the four graphs you can see a linear relationship. The data presented in the third graph down, looks completely random and you will be hard pressed to find any mathematical relationship there.

The correct answer is C.

Question 6

This problem wants you to convert a unit of Millinewtons (mN) to Kilonewtons (kN). Use the conversion ratios to set up the following equations (Newtons will be annotated with an N):

$$500 \text{ mN} \left(\frac{0.001 \text{ N}}{1 \text{ mN}} \right) \left(\frac{1 \text{ kN}}{1000 \text{ N}} \right) = 0.0005 \text{ kN}$$

The correct answer is D.

Question 7

The average number of leaves per lawn is the total number of leaves divided by the number of houses. At each house the following number of leaves were counted:

- House A: 150 leaves

- House B: 300 leaves

- House C: 350 leaves

- House D: 100 leaves

The sum of all the leaves divided by the number of houses, 4, is:

$$\frac{150 + 300 + 350 + 100}{4} = \frac{900}{4} = 225$$

The correct answer is B.

Question 8

In order for this equation to be true the value inside of the absolute value needs to be either a 3 or -3. This is because 3 - 5 = -2 and if either 3 or -3 is inside of the absolute value sign it will come out as a positive 3. You can quickly run through the numbers in the answers to see the when $q = 2$ then $|2 - 5| = 3$ and since 3 - 5 = -2, the correct answer must occur when $q = 2$

The correct answer is C.

Question 9

This question wants you to solve the equation in terms of P. Thee first thing you want to do is subtract the term $\frac{w^2 \rho}{2}$ from both sides of the equation.

$$\frac{v^2 \rho}{2} - \frac{w^2 \rho}{2} = \frac{w^2 \rho}{2} - \frac{w^2 \rho}{2} + P$$
$$\frac{v^2 \rho}{2} - \frac{w^2 \rho}{2} = P$$

Now notice that the factor $\frac{\rho}{2}$ is in both terms on the left side of the equation and can be factored out. Look how that unfolds, only looking at the left side of the equation.

$$\frac{v^2 \rho}{2} - \frac{w^2 \rho}{2} = \frac{\rho}{2}(v^2 - w^2)$$

Which means the new equation looks as follows:

$$\frac{\rho}{2}(v^2 - w^2) = P$$

The correct answer is B.

Question 10

This question is asking for the value of v, not v^2. But, first look to see what the value of v^2 is, because the equation is already solved for it and the values of all the terms on the right side of the equation are given. With the values $w^2 = 50$ and $\frac{P}{\rho} = 25$ you can solve for the value of v^2.

$$v^2 = w^2 + 2\frac{P}{\rho} = 50 + 2(25) = 50 + 50 = 100$$

To solve for v exclusively, take the square root of both sides of the equation.

$$\sqrt{v^2} = \sqrt{100}$$
$$v = 10$$

The correct answer is A.

Question 11

This question is asking you to solve the inequality for y. Begin by multiplying the entire inequality by 2, so that you can rid the of inequality of the fraction.

$$2(\frac{y}{2} + 14 \leq 2y + 8)$$
$$y + 28 \leq 4y + 16$$

Now move all of the y-terms to the right side of the equation by subtracting y from both sides of the equation.

$$y - y + 28 \leq 4y - y + 16$$
$$28 \leq 3y + 16$$

Now move all of the constant terms to the left side of the equation, by subtracting 16 from both sides of the equation.

$$28 - 16 \leq 3y + 16 - 16$$
$$12 \leq 3y$$

To solve for y the final step is to divide the entire equation by 3, in other words multiply the entire equation by $\frac{1}{3}$.

$$\frac{1}{3}(12 \leq 3y)$$
$$\frac{12}{3} \leq \frac{3}{3}y$$
$$4 \leq y$$

Any number that is less than, or equal to, 4 is a solution to this inequality.

The correct answer is D.

Question 12

In order to find the median number of cars sold, order the number of cars sold by each salesman numerically and pick out the middle number. In numerical order the number of cars sold by each salesman is: 2, 4, 8, 14, 16. The middle number is 8.

The correct answer is B.

Question 13

You are looking for a group which makes 11% of the total number of doctors and lawyers. The first step is to find out what 11% of this total is, $193(0.11) = 21.2$. This number is closest to the number of doctors who invest in real estate.

The correct answer is C.

Question 14

The range of a group of numbers is the largest value of the group minus the smallest value of the group. The largest value of this group is 81 and the smallest number is 3. The range is 81 - 3 = 78.

The correct answer is B.

Question 15

The minimum function value of a curve is the y-coordinate for which the curve is at its minimum y-value. This is found on the far-left side of the curve at $x = -5$ where $y = f(x) = -2$.

The correct answer is D.

Question 16

This question wants you to make assumptions based on a best-fit curve of data. It is asking you for the *x*-value (hours of sunlight) of the linear curve when the *y*-value (growth in inches) is 14.5. Go to 14.5 on the vertical axis and move horizontally until you run into the best-fit line. From the line move vertically down until you run into the horizontal axis. Read off the value as 6.

The correct answer is C.

Question 17

To find the slope of the line measure the rise and run between two points. For example to get from the point $(1,12)$ to $(3,13)$ you need to rise by 1 and run by 2, this means the rise over run is $\frac{1}{2}$. This allows only A or B to be an answer to this question. Option A cannot be correct because the line has already moved below 12 on its path to intersect with the vertical axis at $t = 0$. This mean the *g*-intersect must be 11.5.

The correct answer is B.

Question 18

You are looking for an answer which is not a solution to the system of inequalities over the range from $-5 \leq x \leq 5$. The best thing to do is to solve for the outer limits of *y* given the outer limits of *x*. The easiest way is to make a quick table solving each equation for the limits of *x*.

	$x = -5$	$x = 5$
$y \geq 3x+8$	$y \geq -7$	$y \geq 23$
$y \geq 2x+7$	$y \geq -3$	$y \geq 17$

Over the range of $-5 \leq x \leq 5$, the only requirement that needs to be met is that *y* be greater than -7. Whichever answer does not meet this requirement is not a solution.

The correct answer is D.

Question 19

The total money made from all the steel sold, is the price per foot of black steel, *b*, multiplied by the number of feet sold plus the price per foot of stainless steel, *s*, multiplied by the number of feet sold. Mathematically this will look as follows:

$$2.5b + 7.5s = 945$$

This equation only appears in answer B.

The correct answer is B.

Question 20

The 60 dominoes increase by 20% and that total will decrease by 33%. That looks mathematically as follows:

$$60(1.2)(0.667) = 48$$

The correct answer is B.

Question 21

The question is saying that a male student was chosen, that gives you a pool of 70 students to choose from. That 70 will go into the denominator of your probability. The number of male students who eat 2 or less bowls of cereal is $46 + 5 = 51$. This 51 will go into the numerator of your probability.

The correct answer is D.

Question 22

Because every month is shown between January and April, the rate of change of patients transferred to the oncology is the difference in the number of patients in consecutive months. The differences are as follows:

- January and February: $17 - 13 = 4$

- February and March: $13 - 10 = 3$

- March and April: $10 - 7 = 3$

The average rate of change is the average of these three numbers.

$$\frac{4+3+3}{3} = \frac{10}{3} = 3.3$$

The correct answer is D.

Question 23

Look for the largest number in the April column from the four departments in the question. That department is Obstetrics with 23.

The correct answer is D.

Question 24

The equation for a circle is

$$(x-a)^2 + (y-b)^2 = r^2$$

Where:

- a is the x-coordinate of the center point.

- b is the y-coordinate of the center point.

- r is the radius of the circle.

From the equation in the problem $a = 3$, $b = -2$ and $r = 3$ ($9 = 3^2$).

The correct answer is B.

Question 25

This question is asking you to solve for x when F disappears, that is when $F = 0$. With that being the case, the equation looks as follows:

$$0 = -64x + 320$$

Now to solve for x add $64x$ to both sides of the equation.

$$0 + 64x = -64x + 64x + 320$$
$$64x = 320$$

The final step is to divide the entire equation by 64, in other words multiply the entire equation by $\frac{1}{64}$

$$\frac{1}{64}(64x = 320)$$
$$\frac{64}{64}x = \frac{320}{64}$$
$$x = 5$$

The correct answer is A.

Question 26

Katherine runs the mile 10% faster than Richard. In other words, she runs the mile in 90% of the time that it takes Richard.

The correct answer is D.

Question 27

Take a quick look at the number of marbles in each square. Every square has about 25 marbles in it. Being that there are 16 squares, the total number of marbles is 25(16) = 400.

The correct answer is B.

Question 28

A common way to write a parabola is in the following form:

$$y - a = b(x - c)^2$$

Where:

- a is the y-coordinate of the minimum or maximum of the parabola.
- If $b > 0$ the parabola faces up and if $b < 0$ the parabola faces down.
- c is the x-coordinate of the minimum or maximum of the parabola.

Because $b < 0$ in this situation the parabola is facing down, so the maximum will be at the location (a, c). From the equation in this question $a = -1.75$ and $b = 1.5$, this maximum point is located in the fourth quadrant.

The correct answer is D.

Question 29

A common way to write a parabola is in the following form:

$$y - a = b(x - c)^2$$

Where:

- a is the y-coordinate of the minimum or maximum of the parabola.
- If $b > 0$ the parabola faces up and if $b < 0$ the parabola faces down.
- c is the x-coordinate of the minimum or maximum of the parabola.

For this equation you need to move the 2 from the right side of the equation to the left to get it into the proper form. You can accomplish this by adding 2 to both sides of the equation.

$$y + 2 = (x + 4)^2 - 2 + 2$$
$$y + 2 = (x + 4)^2$$

The x and y-coordinates of the vertex are (-4, -2), the sum of these two numbers is -6. This number multiplied by 2 is -12.

The correct answer is A.

Question 30

The equation for a circle is

$$(x - a)^2 + (y - b)^2 = r^2$$

Where:

- a is the x-coordinate of the center point.
- b is the y-coordinate of the center point.
- r is the radius of the circle.

For the circle depicted in this equation the middle point is (3,1), which means $a = 3$ and $b = 1$. The radius needs to be read as the horizontal difference between the points (6,1) and (3,1), which is 6 - 3 = 3. The radius squared is 9.

The correct answer is A.

Question 31

This question is looking for a possible length of time that it will take Geoff to put together 3,000 pizza boxes. Geoff can pack boxes at a rate of between 42 and 78 per minute. The fastest he could pack all of the boxes is 41.667 minutes.

$$\frac{3,000}{72} = 41.667$$

The slowest he could pack all the boxes is 71.43 minutes.

$$\frac{3,000}{42} = 71.43$$

Any number in between would be acceptable.

Question 32

Once the string is long enough that it weighs over 200 pounds it will break under its own weight. It weighs 0.25 pounds per inch and that length for a distance, x, will need to equal 200.

$$0.25x = 200$$

To solve this equation for x divide the entire equation by 0.25, in other words multiply the entire equation by $\frac{1}{0.25}$.

$$\frac{1}{0.25}(0.25x = 200)$$
$$\frac{0.25}{0.25}x = \frac{200}{0.25}$$
$$x = 800$$

The correct answer is 800.

Question 33

The average 10-year-old boy's height, t, multiplied by what number, x, equals the height of a 14-year-old boy, f.

$$tx = f$$

Given the data in the table you can read the following values for t and f.

- $t = 36$

- $f = 51$

Rearrange the equation above so that x is all by itself on the left side of the equation. You can accomplish this by dividing the entire equation by t, in other words multiplying the entire equation by $\frac{1}{t}$.

$$\frac{1}{t}(tx = f)$$
$$\frac{t}{t}x = \frac{f}{t}$$
$$x = \frac{f}{t}$$

Plug the values for f and t into the above equation and solve for the value of x.

$$x = \frac{51}{36} = 1.4$$

The correct answer is 1.4.

Question 34

The first thing you want to do here is find out how many minutes are in 48 hours.

$$48 \text{ hours}\left(\frac{60 \text{ minutes}}{1 hour}\right) = 2,880 \text{ minutes}$$

Now divide the 2,880 minutes by 45 minutes to see how many trips the refrigerator deliver man can make in this time frame.

$$\frac{2,880}{45} = 64$$

If the driver can make 64 trips in 48 hours and deliver 14 refrigerators per trip, then the total number of refrigerators the driver can deliver in this period is $14(64) = 896$.

The correct answer is 896.

Question 35

The volume of a sphere is:

$$V = \frac{4}{3}\pi r^3$$

Since you are looking for the diameter, you will first want to solve for the radius, r. Actually, it will be best to first solve r^3 for its numerical value and then take the cubed root of that. To solve for r^3 multiply the entire equation by $\frac{3}{4}$.

$$\frac{3}{4}\left(V = \frac{4}{3}\pi r^3\right)$$
$$\frac{3}{4}V = \pi r^3$$

Now divide the entire equation by π, in other words multiply the entire equation by $\frac{1}{\pi}$.

$$\frac{1}{\pi}\left(\frac{3}{4}V = \pi r^3\right)$$
$$\left(\frac{3}{4}\right)\frac{V}{\pi} = r^3$$

Plug the value 288π into V and solve r^3.

$$r^3 = \left(\frac{3}{4}\right)\frac{288\pi}{\pi} = \left(\frac{3}{4}\right)288 = 216$$

In order to solve for r take the cubed root of $r^3 = 384$.

$$(r^3 = 216)^{\frac{1}{3}}$$
$$r = \sqrt[3]{216}$$
$$r = 6$$

The question is asking for the diameter, which is twice the radius of 6.

The correct answer is 12.

206

Question 36

A function, equation or fraction is undefined when the denominator is 0. This knowledge turns the problem into solving a quadratic equation.

$$t^2 + 5t + 6$$

Before jumping into the quadratic formula to solve this problem, try and factor it out by finding two numbers which add up to 5 and multiply to 6. Those two numbers are 2 and 3. Which means the above equation factors to:

$$(t + 2)(t + 3)$$

Set the factored form of the polynomial equal to 0 so that it will render the denominator 0 when solved for t.

$$(t + 2)(t + 3) = 0$$

Setting either one of these sums equal to 0 will solve the equation, so that the first possible solution looks as follows:

$$t + 2 = 0$$
$$t + 2 - 2 = 0 - 2$$
$$t = -2$$

The second possible solution looks as follows:

$$t + 3 = 0$$
$$t + 3 - 3 = 0 - 3$$
$$t = -3$$

The correct answers are -2 or -3.

Question 37

When the bacteria grow the number 365 will increase, this means it will have to be multiplied by a number larger than 1. The number 12% written in decimal form is 0.12. So, if you want the bacteria to grow by 12%, multiply the 365 by 1 with a 0.12 added to it, that number is 1.12.

The correct answer is 1.12.

Question 38

First calculate how many bacteria will be present in the farm which started with 400 bacteria after 14 days.

$$400(1.12)^{14} = 1954.84$$

Then calculate the number of bacteria after 14 days for the bacteria farm which started with 365 bacteria.

$$365(1.12)^{14} = 1783.80$$

You are looking for how many more bacteria the farm, which started with 400 bacteria, has than the one with 365. That is the difference between 1954.84 - 1783.8 = 171.

The correct answer is 171.

Chapter 9

Practice Test 3

Practice Math Test - No Calculators Allowed

Directions

You will have 25 minutes to complete part 3 of the math practice test. There are 20 questions for this portion of the test. Questions 1-15 are multiple choice and questions 16-20 are bubble-in questions where no choices are provided for you. You may use a piece of scratch paper for notes and calculations. Write your answers down on a separate piece of paper, the solutions and explanations can be found immediately after the test. You are to continue when you see the word CONTINUE in the bottom right corner of the page. Stop taking the test when the word STOP appears in the bottom center of the page. If you finish before the time expires you may check your answers for part 3 of the test only, do not move ahead within the 25 minutes provided for the portion of the test.

Notes

- No calculator usage allowed

- All variables and expressions belong to the set of real numbers unless otherwise written.

- Not all figures are drawn to scale. If the scaling is important it will be noted.

- If a drawing is in three-dimensions it will be noted, otherwise all drawings are in two-dimensions.

CONTINUE

Reference Equations

- The area of a circle is $A = \pi r^2$

- The circumference of a circle is $C = 2\pi r$

    r is the radius of the circle

- The area of a rectangle is $A = lw$

    l is the length of the rectangle

    w is the width of the rectangle

- The area of a triangle is $A = \frac{1}{2}bh$

    b is the base of the triangle

    h is the height of the triangle

- The sides of a right triangle have the following relationship: $a^2 + b^2 = c^2$

    c is the length of the hypotenuse

    a and b are the side lengths for the sides whose ends meet at the right angle

- A right triangle with angles of $30°$, $60°$ and $90°$ angles has side lengths of x and $x\sqrt{3}$ and a hypotenuse length of $2x$.

- A right triangle with angles of $45°$, $45°$ and $90°$ angles has side lengths of x and x and a hypotenuse length of $2x\sqrt{2}$.

- The volume of a rectangular prism is $V = lwh$

    h is the height of the prism

- The volume of a cylinder is $V = \pi r^2 h$

- The volume of a sphere is $V = \frac{4}{3}\pi r^3$

- The volume of a cone is $V = \frac{1}{3}\pi r^2 h$

- The volume of a pyramid is $V = \frac{1}{3}lwh$

- There are $360°$ in the arc of a circle

- There are 2π radians in the arc of a circle

- The sum of all the angles in a triangle equals $180°$

CONTINUE

1.

Which of the following inequalities can never be true?

A) $|-2x| > 0$

B) $|x - 50| > 0$

C) $|3x + 25| < 0$

D) $|-\frac{x}{2} - 15| > 0$

2.

$$g(y) = 2y + k$$

For the function $g(y)$ above, when $y = -2$ then $g(y)$ equals 2. What is $g(y)$ when $y = 3$?

A) 12

B) 2

C) -10

D) -12

3.

$$xy = 7$$
$$y + \frac{1}{x} = 1$$

If (x, y) is a solution to the system of equations above, what is the value of x?

A) 7

B) 8

C) 9

D) 10

4.

If $g(y) = 13 + 21y$, what is $g(\frac{1}{3}y - \frac{1}{7})$?

A) $7y + 16$

B) $-7y + 16$

C) $-7y - 10$

D) $7y + 10$

CONTINUE

5.

$$\frac{2(x+3)(4x-1)}{8x-2}$$

Which of the following is equivalent to the expression above?

A) $\dfrac{4x^2 + 11x + 3}{4x - 1}$

B) $x + 3$

C) $\dfrac{8x^2 + 22x - 6}{4x - 1}$

D) $2(x + 3)$

6.

$$\frac{3}{10}x^{1/4} = \frac{3 - \sqrt{x}}{\sqrt[4]{x}}$$

How can the above equation be rewritten?

A) $\dfrac{13}{10} = \dfrac{3}{\sqrt{x}}$

B) $\dfrac{3}{10}\sqrt[4]{x} = 3 - \sqrt{x}$

C) $\dfrac{3}{10} = \dfrac{3 - \sqrt{x}}{\sqrt[8]{x}}$

D) Answers A - C are all incorrect.

7.

Tommy is training for a nail hammering competition, between him and some of the neighborhood kids, being held in four weeks. He plans on decreasing the time it takes him to hammer in a nail by 1.5 seconds every day until the competition. Which of the following statements about Tommy cutting down his time is true?

A) Tommy will decrease his hammering time by 4.5 seconds every 4 days.

B) Tommy will increase his hammering time by 3 seconds every 2 days.

C) Tommy will decrease his time by 42 seconds in total before the competition.

D) Tommy will increase his time by 42 seconds in total before the competition.

CONTINUE

8.

Which of the following lines is perpendicular to the line $s = 2t + 3$?

A) $2s = t + 3$

B) $s = \dfrac{1}{2}t + 3$

C) $s - 2t = -3$

D) $s + \dfrac{1}{2}t = 3$

9.

$$\frac{4}{\sqrt{b-a}} = 2$$

If $b = 5$, what is the value of a in the equation above?

A) -1

B) 0

C) 1

D) 2

10.

If $\dfrac{x+6}{2-x} = 7$, what is the value of x?

A) $\dfrac{9}{20}$

B) $-\dfrac{9}{20}$

C) -1

D) 1

11.

From the equation $\dfrac{x^a}{x^b} = x^{12}$, if $b = -2$ what is the value of a?

A) 10

B) 14

C) -6

D) Not enough information to solve this problem.

CONTINUE

12.

Kevin and Kasey spent three weeks studying for the same test. Kevin spent 30% more time studying than Kasey and the combined time spent studying was 230 hours. How much time did Kevin spend studying?

A) 100

B) 50

C) 130

D) 65

13.

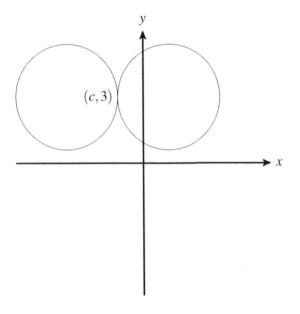

The equations for the circles above are $(x-2)^2 + (y-3)^2 = 16$ and $(x+6)^2 + (y-3)^2 = 16$. The two circles intersect at the point $(c, 3)$. Which of the following is the value of c?

A) -4

B) -3

C) -2

D) -1

14.

$$\frac{3+2i}{2-4i}$$

How can the complex number above be rewritten in the form $a + bi$? (Note: $i = \sqrt{-1}$)

A) $3 + 2i$

B) $2 - 4i$

C) $0.1 - 0.8i$

D) $-0.1 + 0.8i$

15.

$$-y^2 + dy = -c$$

In the quadratic equation above, both d and c are constants. What is the solution for y?

A) $\dfrac{-d \pm \sqrt{d^2 - 4c}}{2}$

B) $\dfrac{d \pm \sqrt{d^2 + 4c}}{2}$

C) $\dfrac{d \pm \sqrt{d^2 + 4c}}{2d}$

D) $\dfrac{d \pm \sqrt{d^2 - 4c}}{2}$

CONTINUE

Directions

The final questions in part 3 of the test are to be bubbled in. On the actual test there will be pictures to show you how to properly bubble in the answers on your answer sheet. For this book, you only need to write down your answers on a separate sheet of paper. You will be able to bubble in fractions and decimals on the actual test, so don't worry if answers appear in this form on the practice test.

CONTINUE

16.

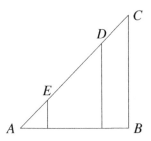

In the right triangle above the distances $\overline{AB}$ and $\overline{BC}$ are 4 and 3, the distances $\overline{AE}$, $\overline{ED}$ and $\overline{DC}$ are s, $2s$ and s. What is the value of $2s$?

17.

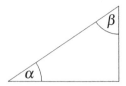

In the triangle above the value of $\cos(\alpha) = 0.7$, what is the value of $\sin(\beta)$?

18.

$$-2x^4 + 3x^3 - 14x^2 + 5x + 12 = 2a \qquad (9.1)$$

In the equation above $a = 6$, for what value of x is the above equation true?

CONTINUE

19.

$$7q + 4r = 18$$
$$2q - r = 3$$

If (q, r) is a solution to the systems of equations of above. What is the value of r?

20.

Julie is a saleswoman at a large washing machine company. She receives a base salary of 100 dollars a month along with 20 dollars for every washing machine sold. If in July Julie sold 13 washing machines, how much money did she make total in that month?

STOP

If you finished early, check only your work for this section.
Do not look at any other section.

Directions

You will have 55 minutes to complete part 4 of the math practice test. There are 38 question on this portion of the test. Questions 1-30 are multiple choice and questions 31-38 are bubble-in questions where no choices are provided for you. You may use a piece of scratch paper for notes and calculations. Write your answers down on a separate piece of paper, the solutions and explanations can be found immediately after the test. You are to continue when you see the word CONTINUE in the bottom right corner of the page. Stop taking the test when the word STOP appears in the bottom center of the page. If you finish before the time expires you may check your answers for part 4 of the test only, do not return to part 3 within the 55 minutes provided for the portion of the test.

Notes

- Calculator usage is allowed

- All variables and expressions belong to the set of real numbers unless otherwise written.

- Not all figures are drawn to scale. If the scaling is important it will be noted.

- If a drawing is in three-dimensions it will be noted, otherwise all drawings are in two-dimensions.

CONTINUE

Reference Equations

- The area of a circle is $A = \pi r^2$

- The circumference of a circle is $C = 2\pi r$

    r is the radius of the circle

- The area of a rectangle is $A = lw$

    l is the length of the rectangle

    w is the width of the rectangle

- The area of a triangle is $A = \dfrac{1}{2}bh$

    b is the base of the triangle

    h is the height of the triangle

- The sides of a right triangle have the following relationship: $a^2 + b^2 = c^2$

    c is the length of the hypotenuse

    a and b are the side lengths for the sides whose ends meet at the right angle

- A right triangle with angles of $30°$, $60°$ and $90°$ angles has side lengths of x and $x\sqrt{3}$ and a hypotenuse length of $2x$.

- A right triangle with angles of $45°$, $45°$ and $90°$ angles has side lengths of x and x and a hypotenuse length of $2x\sqrt{2}$.

- The volume of a rectangular prism is $V = lwh$

    h is the height of the prism

- The volume of a cylinder is $V = \pi r^2 h$

- The volume of a sphere is $V = \dfrac{4}{3}\pi r^3$

- The volume of a cone is $V = \dfrac{1}{3}\pi r^2 h$

- The volume of a pyramid is $V = \dfrac{1}{3}lwh$

- There are $360°$ in the arc of a circle

- There are 2π radians in the arc of a circle

- The sum of all the angles in a triangle equals $180°$

CONTINUE

1.

$$t^3 + 2t^2 - 16t - 32 = 0$$

Which of the following is a possible solution to the equation above?

A) -3

B) -4

C) -5

D) -6

2.

Jason digs holes for a living. He can dig 300 holes every 8 hours. If he shows up to a job site and there are already 78 holes dug, which of the following equations can be used to estimate the number of hours, h, he will need to dig holes until 248 holes are dug?

A) $248 = 78 + \dfrac{h}{300}$

B) $78 = 248 + 300h$

C) $248 = 78 + 37.5h$

D) $248 = 78 + \dfrac{h}{37.5}$

3.

A group of 30 people were put into a room to be separated into smaller groups for team building activities. The entire group was separated into thirds and those groups were separated into fifths. How many people were in each group after the two separations?

A) 30

B) 15

C) 10

D) 2

4.

A group of 638 lawyers were surveyed on whether they would rather use a passive or an active voice when writing their legal documents. Of those surveyed 68.5% said they preferred using an active over a passive voice. What is the number of lawyers, from those surveyed who preferred using a passive voice?

A) 201

B) 437

C) 420

D) 218

CONTINUE

5.

The momentum of an object equals its mass multiplied by its velocity. If the momentum of an object is 16 and the velocity of an object is 4, what is the mass of the object?

A) 1

B) 2

C) 4

D) 8

6.

Jane and Susy ate pizza exclusively over a two-week period. Susy ate 14 more slices of pizza than Jane. In total they ate 178 slices of pizza. If each pizza they ate had 7 slices, how many more complete pizzas did Susy eat?

A) 82

B) 96

C) 12

D) 13

7.

	1980's	1990's	2000's	2010's
Regular Coffee	10	16	17	5
Decaff Coffee	24	2	14	23
Cappuccino	14	3	11	17
Mocha	3	23	8	22
Latte	21	22	21	5

The above table shows the average number of 5 different types of coffee sold per day at a local coffee shop over the decades from the 1980s to the 2010s. From the data presented what percentage of coffee's sold per day in the 1990s were Mochas?

A) 41%

B) 35%

C) 4%

D) 11%

8.

The line $y = \frac{1}{2}x - 3$ lies on the xy-plane. It passes through 3 of the 4 quadrants on the plane. Which quadrant does the line not pass through?

A) I

B) II

C) III

D) IV

CONTINUE

220

9.

	Cows	Chickens	Sheep	Horses	Goats	Pigs	Total
Texas	466	489	466	522	407	572	2,922
Nebraska	533	409	448	473	410	450	2,723
Kentucky	597	522	405	402	501	413	2,840
Tennessee	442	468	513	449	506	530	2,908
Missouri	495	485	408	417	517	462	2,784
Total	2,533	2,373	2,240	2,263	2,341	2,427	14,177

The table above shows the numbers of farm animals per 100 square miles in 5 American states. If an animal was chosen at random from the table, what is the probability that it would be a pig from Tennessee or Kentucky?

A) 0.389

B) 0.324

C) 0.336

D) 0.067

CONTINUE

The following information is for questions 10 and 11.

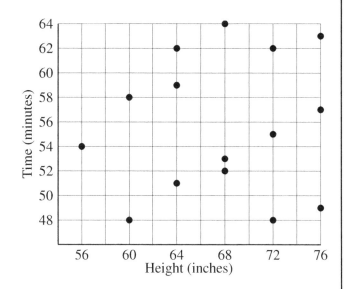

Height (inches)

The graphic above shows race times by people of different heights.

10.

What is the fastest race time of a person who is 72 inches tall?

A) 48

B) 55

C) 64

D) 52

11.

What is the median race time of a person who is 64 inches tall?

A) 51

B) 59

C) 62

D) 48

12.

The function $P(x)$ when plotted in the xy-plane has x-intercepts at 5, 2 and -3. Which of the following is a possible option for the function $P(x)$?

A) $P(x) = (x+5)(x+2)(x-3)$

B) $P(x) = (x+5)^2(x-3)$

C) $P(x) = (x-5)^2(x+3)$

D) $P(x) = (x-5)(x-2)(x+3)$

CONTINUE

13.

Year	Dollars
0	25
1	30
2	35
3	40
4	45
5	50

According to the table above, in year one Larry deposited $25 in a bank account. The table shows the new dollar amount at the start of each new year. How is Larry's money changing over time?

A) increasing exponentially

B) decreasing exponentially

C) increasing linearly

D) decreasing linearly

14.

$$w(x) = \frac{Mx^2}{2EI}$$

The above equation shows the displacement for a cantilevered beam with a moment acting on its end. M is the moment acting at the end of the beam, x is the location where you are analyzing the displacement of the beam, E is material specific Young's modulus, and I is the moment of inertia based on the cross section of the beam. If the location being analyzed, x, is doubled how does the displacement w change?

A) The displacement stays the same

B) The displacement is doubled

C) The displacement is tripled

D) The displacement is quadrupled

CONTINUE

15.

Which of the following is an example of exponential growth?

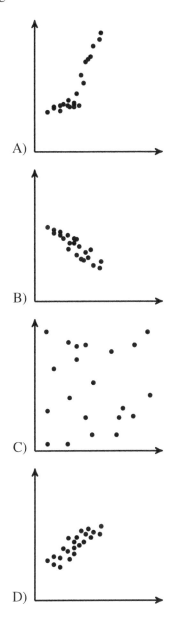

A)

B)

C)

D)

The following information is for questions 16 and 17
Grape company X is looking to put in a new irrigation system for their hundreds of acres of vineyards. They look at the prices of three different irrigation companies to try to find the best deal. All prices in the table are per acre.

	Design Cost	Material Cost	Labor Cost
Company 1	17	700	300
Company 2	13	600	250
Company 3	25	500	400

The total cost of an irrigation system, I, is $I = (D + M + L)A$. Where M is the material cost, D is the design cost, L is the labor cost and A is the number of acres.

16.

Grape company X has a budget of $100,000 for an irrigation system this year. How many acres could they irrigate if they went with Company 3?

A) 115

B) 108

C) 98

D) 129

CONTINUE

17.

If Company 1 offers a rebate of $20,000 with a purchase of an irrigation system. How many acres would Grape Company X need to irrigate in order for it to be cheaper to buy an irrigation system from Company 2?

A) 218

B) 117

C) 92

D) 130

18.

Stacey has spherical containers, each with a radius of 2 inches. If Stacey has 14 gallons of putty that she wishes to put into the spherical containers. How many complete containers can she fill with the putty? (Note: 1 gallon equals 231 cubic inches)

A) 76

B) 86

C) 96

D) 106

19.

If $7q + 5 \leq 6$, what is the maximum of $7q - 5$?

A) 6

B) 5

C) -5

D) -4

CONTINUE

20.

The population of a city, P, has shown that it triples every year. If the initial population of the city is 10,000 people and the years are expressed by the variable t, which of the following equations expresses the population growth?

A) $P = 10000(3)t$

B) $P = \dfrac{10000}{3}t$

C) $P = 10000(3)^t$

D) $P = 3(10000)^t$

The following information refers to questions 21 and 22.

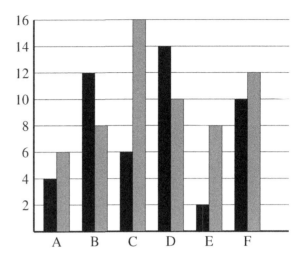

The bar graph above shows the sales in millions of 6 different products in the first and second quarters of the fiscal year. The black highlighted bars represent sales in the first quarter, and the grey highlighted bars represents the sales in the second quarter.

21.

Which product shows the largest increase in sales between the first and second quarters of the fiscal year?

A) Product B

B) Product C

C) Product D

D) Product E

CONTINUE

22.

If the sales of Product E are growing at a linear rate, how many sales can be expected in the third quarter?

A) 12 million

B) 14 million

C) 16 million

D) 18 million

23.

			Power Supply 1		
Voltage	22	23	24	25	26
Frequency	10	8	3	7	9

			Power Supply 2		
Voltage	10	11	12	13	14
Frequency	1	9	14	10	3

The voltages for 37 24-Volt household devices and 12-Volt household devices were measured. The measured voltages were tabulated above. Which of the following is true about the two data sets?

A) The standard deviation for Power Supply 1 is larger.

B) The standard deviation for Power Supply 2 is larger.

C) The standard deviations for both power supplies are the same.

D) There is not enough information to determine the standard deviation of the two power supplies.

CONTINUE

24.

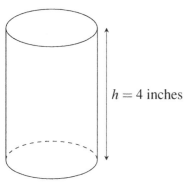

$h = 4$ inches

The cylinder above has a volume of 16π cubic inches. Which of the following is the circumference of the circular portion of the cylinder?

A) 2

B) 4

C) 2π

D) 4π

25.

$$g(t) = 6t^3 + 15t^2 + 9t$$
$$h(t) = 2t^2 + 5t + 3$$

The polynomials $g(t)$ and $h(t)$ are defined above. Which of the following polynomials is divisible by $(6t - 4)$?

A) $P(t) = g(t) + h(t)$

B) $P(t) = g(t) - 4h(t)$

C) $P(t) = 2g(t) - 4h(t)$

D) $P(t) = g(t) - 4h(t)$

26.

If a and b are numbers so that $|a + 3| \leq b$, which of the following must be true?

   I. $-|a + 3| \geq -b$

  II. $a + 3 \leq b$ (with $a \geq -3$)

 III. $a + 3 \leq b$ (with $a \leq -3$)

A) I only

B) I and II

C) I, II and III

D) I and III

CONTINUE

27.

The price per kilometer traveled by train in the German state of Bavaria decreases with the number of kilometers a person plans on traveling each month. The prices may also vary depending on in what month the tickets are bought.

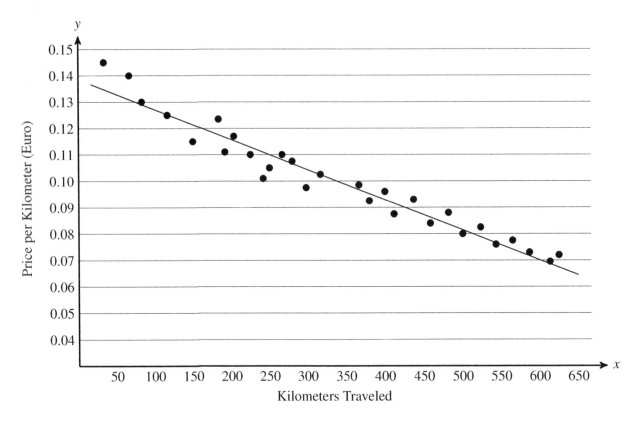

The actual prices paid for train tickets per kilometer traveled are shown in the scatter plot above. The equation of the best fit line for this scatter plot is $y = 0.1385 - 1.14 \times 10^{-4}x$ and is shown on the graph above. Which of the following best explains the value 1.14×10^{-4} shown in the best fit equation?

A) The expected payment for every kilometer traveled is 1.14×10^{-4} Euros.

B) The price per kilometer increases by 1.14×10^{-4} Euros for every kilometer traveled.

C) The price per kilometer decreases by 1.14×10^{-4} Euros for every kilometer traveled.

D) The price per kilometer traveled will never exceed 1.14×10^{-4} Euros.

CONTINUE

28.

$$g(y) = (y+3)^2 - 25$$

The vertex form of the function $g(x)$ is shown above. How can this equation be rewritten so that the function is in terms of the y-intercept?

A) $g(y) = (y-3)^2 - 25$

B) $g(y) = (y+8)(y-2)$

C) $g(y) = y^2 + 6y - 16$

D) $g(y) = (y-8)(y+2)$

29.

If q is the average of $2a, 3b$ and 56, r is the average of $4a, 9b$ and 17, and s is the average of $6a, 6b$ and 59, what is the average of $q, r,$ and s in terms of a and b?

A) $4a + 3b + 22$

B) $4a + 6b + 44$

C) $6a + 9b + 66$

D) $2a + 3b + 132$

30.

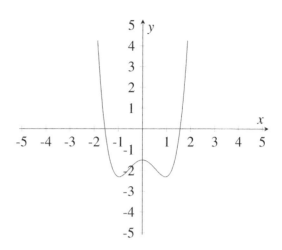

A math teacher is looking to build a system of equations with the most possible solutions. The first equation he chose for his system of equations was $y = x^4 - 1.8x^2 - 1.5$, this equation is shown on the graph above. Which of the following equations will produce the most possible solutions to the system of equations created by the teacher, when paired with the equation graphed above?

A) $y = 2x - 2$

B) $y = \dfrac{1}{4}x - 2$

C) $y = -2$

D) $y = \sqrt{4 - x^2}$

CONTINUE

Directions

The final questions in part 4 of the test are to be bubbled in. On the actual test there will be pictures to show you how to properly bubble in the answers on your answer sheet. For this book, you only need to write down your answers on a separate sheet of paper. You will be able to bubble in fractions and decimals on the actual test, so don't worry if answers appear in this form on the practice test.

CONTINUE

31.

A gumball machine is full of 336 gumballs. If the machine can empty at a rate of 12 gumballs per minute, how many seconds will it take to empty exactly half of the gumballs from the machine?

32.

The gas, G, required to lift a hot air balloon increases for every person, p, who is in the basket during the flight. The gas requirement can be modeled using the equation $G = \dfrac{2p + 13}{5}$, how much more gas will a hot air balloon need for each additional passenger in the balloon?

33.

Mark spilled orange and green beads on his kitchen table. He noticed that there were 180 more orange beads than there were green beads. He also noticed that exactly $\dfrac{7}{9}$ of the beads were orange. How many green beads were on the table?

34.

The Martian race uses the unit *tic* to measure distance. A *tic* is made up of exactly 17 *tacs* and is equivalent to 63.47 centimeters. How many *tacs* are in 2.5 meters? Round your answer to the nearest tenth. (Note: 100 centimeters = 1 meter)

CONTINUE

35.

$$I = \frac{\pi d^4}{32}$$

The above equation describes the relationship between the moment of inertia of a circular rod and its diameter. What is the ratio of a moment of inertia with a diameter, d, to a moment of inertia with a diameter $2.5d$? Round to the nearest thousandth.

36.

What is the arc length of three quarters of the circumference of the unit circle? Round to the nearest hundredth.

The following information is for both questions 37 and 38.

The current elevation of a city in the California central valley is 14 feet above sea level. Experts believe that the city will lose 1% of its elevation every year due to continued use of the water reserves underneath the city. A model for the elevation of the city looks as follows:

$$E = 14(r)^t$$

37.

Given that the city is losing elevation by 1% per year, what should the value of r be?

38.

What will be the elevation of the city after 8 years? Round to the nearest foot.

STOP
If you finished early, check only your work for this section.
Do not look at any other section.

9.1 Practice Test 3 Solutions

Part 3: Multiple Choice

1. C
2. A
3. B
4. D
5. B
6. A
7. C
8. D
9. C
10. D
11. A
12. C
13. C
14. D
15. B

Part 3: Bubble In

16. 2.5
17. 0.7
18. 0
19. 1
20. 360

Part 4: Multiple Choice

1. B
2. C
3. D
4. A
5. C
6. D
7. B
8. A
9. D
10. A
11. B
12. D
13. C
14. D
15. A
16. B
17. D
18. C
19. D
20. C
21. B
22. B
23. A
24. D
25. C
26. B
27. C
28. B
29. B
30. D

Part 4: Bubble In

31. 840
32. $\frac{2}{5}$
33. 72
34. 67
35. 0.026
36. 4.71
37. 0.99
38. 13

9.2 Practice Test 3 Explanations

Question 1

An absolute value will always render a positive number. Because the absolute value in option C is shown as being less than 0, this cannot be true.

The correct answer is C.

Question 2

Because the value of $g(y)$ when $y = -2$ is given as 2, you can solve for the value of k. First rearrange the function to solve it for k, that is accomplished by subtracting $2y$ from both sides of the equation.

$$g(y) - 2y = 2y - 2y + k$$
$$g(y) - 2y = k$$

Now plug in the value $g(y)$ when $y = -2$ and set y to -2, then solve for k.

$$2 - 2(-2) = k$$
$$2 + 4 = k$$
$$6 = k$$

Now plug in the value $y = 3$ into the original function $g(y)$ to find the function value.

$$g(y) = 2(3) + 6 = 12$$

The correct answer is A.

Question 3

Because you are solving only for the value x, you will want to solve first for the variable y from either equation, and then plug that y-value into the other equation to solve for x. Begin with solving the bottom equation for y by subtracting $\dfrac{1}{x}$ from both sides of the equation.

$$y + \frac{1}{x} - \frac{1}{x} = 1 - \frac{1}{x}$$
$$y = 1 - \frac{1}{x}$$

Plug this value for y into the top equation.

$$x\left(1 - \frac{1}{x}\right) = 7$$

Distribute the x over the sum in the parenthesis.

$$x - \frac{x}{x} = 7$$
$$x - 1 = 7$$

To solve for x add 1 to both sides of the equation.

$$x - 1 + 1 = 7 + 1$$
$$x = 8$$

The correct answer is B.

Question 4

In this equation the value to be plugged into the function $g(y)$ in the place of y is $\frac{1}{3}y - \frac{1}{7}$.

$$g\left(\frac{1}{3}y - \frac{1}{7}\right) = 13 + 21\left(\frac{1}{3}y - \frac{1}{7}\right)$$

Distribute the 21 over the sum in the parenthesis.

$$g\left(\frac{1}{3}y - \frac{1}{7}\right) = 13 + \frac{21}{3}y - \frac{21}{7} = 13 + 7y - 3$$

At this point combine the constant terms.

$$g\left(\frac{1}{3}y - \frac{1}{7}\right) = 10 + 7y = 7y + 10$$

The correct answer is D.

Question 5

This question is testing your ability to notice when to factor out numbers in order to cancel out terms in a fraction. Take the fraction and factor out a 2 from the sum in the denominator.

$$\frac{2(x+3)(4x-1)}{8x-2} = \frac{2(x+3)(4x-1)}{2(4x-1)}$$

Notice that the 2 and the sum $(4x-1)$ cancel out from the numerator and the denominator, after the cancellation only $x + 3$ will remain.

The correct answer is B.

Question 6

First thing to remember is that $\sqrt[4]{x} = x^{\frac{1}{4}}$. With this knowledge multiply the entire equation by $x^{\frac{1}{4}}$, this will cancel out the denominator on the right side of the equation.

$$x^{1/4}\left(\frac{3}{10}x^{1/4} = \frac{3-\sqrt{x}}{\sqrt[4]{x}}\right)$$
$$\frac{3}{10}x^{1/4}x^{1/4} = \frac{x^{1/4}}{x^{1/4}}(3-\sqrt{x})$$
$$\frac{3}{10}x^{1/4}x^{1/4} = 3-\sqrt{x}$$

Now refer to rule 1 from section 2.5 in chapter 2. Notice that $x^{1/4}x^{1/4} = x^{1/4+1/4}$ and this equals $x^{1/2}$ which in turn equals $\sqrt{x}$. All this is to say that the equation above can be rewritten as follows:

$$\frac{3}{10}x^{1/4}x^{1/4} = \frac{3}{10}\sqrt{x} = 3 - \sqrt{x}$$

Now divide the entire equation by $\sqrt{x}$, in other words multiply the entire equation by $\frac{1}{\sqrt{x}}$.

$$\frac{1}{\sqrt{x}}\left(\frac{3}{10}\sqrt{x} = 3 - \sqrt{x}\right)$$

$$\frac{3}{10}\frac{\sqrt{x}}{\sqrt{x}} = \frac{3 - \sqrt{x}}{\sqrt{x}}$$

$$\frac{3}{10} = \frac{3 - \sqrt{x}}{\sqrt{x}}$$

Now separate the fraction on the right side of the equation into two fractions and see what cancels out.

$$\frac{3}{10} = \frac{3}{\sqrt{x}} - \frac{\sqrt{x}}{\sqrt{x}}$$

$$\frac{3}{10} = \frac{3}{\sqrt{x}} - 1$$

Move the 1 onto the left side of the equation and rewrite it as the fraction $\frac{10}{10}$.

$$\frac{3}{10} + 1 = \frac{3}{\sqrt{x}} - 1 + 1$$

$$\frac{3}{10} + 1 = \frac{3}{\sqrt{x}}$$

$$\frac{3}{10} + \frac{10}{10} = \frac{3}{\sqrt{x}}$$

Add the $\frac{3}{10}$ with the $\frac{10}{10}$ to get $\frac{13}{10}$.

$$\frac{3}{10} + \frac{10}{10} = \frac{3}{\sqrt{x}}$$

$$\frac{13}{10} = \frac{3}{\sqrt{x}}$$

The correct answer is A.

Question 7

If Tommy is decreasing his time by 1.5 seconds every day, after 4 days he will decrease his time by (4)1.5 = 6 seconds. Choice A is incorrect. Choice B is also incorrect because Tommy does not want to increase his time but decrease it. For the same reason choice D is also incorrect. After four weeks or 28 days Tommy will decrease his hammering time by (28)1.5 = 42 seconds.

The correct answer is C.

Question 8

In order for two lines to perpendicular to one another the product of their slopes must be equal to -1. Which means you need to set up and solve the following equation:

$$2x = -1$$

Where 2 is the slope of the line in the question and x is the slope of the perpendicular line that we are trying to solve for. To solve this equation for x divide the entire equation by 2, in other words multiply the entire equation by $\frac{1}{2}$.

$$\frac{1}{2}(2x = -1)$$
$$x = -\frac{1}{2}$$

Looking at the answers there is no equation with this slope present; however only two equations have a slope of $\frac{1}{2}$. Choice B cannot be correct because it is positive and on the opposite side of the equation from the variable s. If you subtract $\frac{1}{2}t$ from both sides of the equation for choice D, then the following becomes apparent:

$$s + \frac{1}{2}t - \frac{1}{2}t = -\frac{1}{2}t + 3$$
$$s = = -\frac{1}{2}t + 3$$

The correct answer is D.

Question 9

This question is basically asking you to solve the equation for a. Start by multiplying the entire equation by $\sqrt{b-a}$, this will cancel out the $\sqrt{b-a}$ term in the denominator of the fraction on the left side of the equation.

$$\sqrt{b-a}\left(\frac{4}{\sqrt{b-a}}\right) = 2$$
$$4 = 2\sqrt{b-a}$$

The next step is to divide the entire equation by 2, in other words multiply the entire equation by $\frac{1}{2}$.

$$\frac{1}{2}(4 = 2\sqrt{b-a})$$
$$\frac{4}{2} = \frac{2}{2}\sqrt{b-a}$$
$$2 = \sqrt{b-a}$$

The next thing to do is to get rid of the square root; you can do that by squaring both sides of the equation.

$$(2)^2 = (\sqrt{b-a})^2$$
$$4 = b - a$$

238

Since we are trying to solve for a and not negative a at this point you can bring the a to the left side of the equation. That can be accomplished by adding a to both sides of the equation. While you are at it, subtract 4 from both sides of the equation.

$$4-4+a=b-a+a-4$$
$$a=b-4$$

When the value of b is 5 then $a = 5 - 4 = 1$.

The correct answer is C.

Question 10

To solve this equation for x you need to get all the x-terms out of any denominator. Begin by multiplying the entire equation by $2-x$, this will cancel out the $2-x$ term from the denominator on the left side of the equation.

$$(2-x)\left(\frac{x+6}{2-x}=7\right)$$
$$x+6=7(2-x)$$

Now distribute the 7 over the sum in the parenthesis on the right side of the equation.

$$x+6=7(2-x)$$
$$x+6=14-7x$$

Move all of the x-terms over to the right side of the equation by adding $7x$ to both sides of the equation.

$$x+7x+6=14-7x+7x$$
$$8x+6=14$$

Move all the constant terms to the right side of the equation by subtracting 6 from both sides of the equation.

$$8x+6-6=14-6$$
$$8x8$$

To solve this equation exclusively for x, divide the entire equation by 8, in other words multiply the entire equation by $\frac{1}{8}$.

$$\frac{1}{8}(8x=8)$$
$$\frac{8}{8}x=\frac{8}{8}$$
$$x=1$$

The correct answer is D.

Question 11

Refer back to the rules for exponents in section 2.5 of chapter 2. The left side of the equation can be rewritten as follows:

$$\frac{x^a}{x^b} = x^{a-b}$$

So that the equation from the question can be rewritten as follows:

$$x^{a-b} = x^{12}$$

Because the expressions on the left and right side of the equation both have x as the base of the exponents, the following must be true:

$$a - b = 12$$

Because the value of b is given you can solve for the value of a at this point. This can be accomplished by adding b to both sides of the equation.

$$a - b + b = 12 + b$$
$$a = 12 + b$$

Plug in the value of $b = -2$ into the equation and solve for the numerical value of a.

$$a = 12 + (-2) = 12 - 2 = 10$$

The correct answer is A.

Question 12

Call the time the Kevin spent studying V and the time Kasey spent studying K. If Kevin spent 30% more time studying than Kasey, then you can write down the following equation:

$$V = 1.3K$$

The total combined time spent studying was 230 hours, which means the following equation must also be true:

$$V + K = 230$$

Because you are trying to solve for how long Kevin spent studying, you will need to solve for V. Usually I would recommend solving for V first; however in this situation that will turn out to be quite cumbersome, try it out to see, but go ahead and begin by solving for K first by taking the value for V from the first equation and plugging it into the second equation.

$$V + K = 230$$
$$1.3K + K = 230$$

Now you can factor out the K on the left side of the equation.

$$1.3K + K = 230$$
$$K(1.3 + 1) = 230$$
$$K(2.3) = 230$$

To solve for K exclusively divide the entire equation by 2.3, in other words multiply the entire equation by $\frac{1}{2.3}$.

$$\frac{1}{2.3}(K(2.3) = 230)$$
$$K = \frac{230}{2.3}$$
$$K = 100$$

The value for K you can plug into the first equation $V = 1.3K$ to solve for the number of hours Kevin spent studying.

$$V = 1.3K = (1.3)100 = 130$$

The correct answer is C.

Question 13

To solve this question, you will need to figure out the x-coordinate of the center point of one of the circles and the radius of that circle. Look at the first equation of a circle given in the problem.

$$(x-2)^2 + (y-3)^2 = 16$$

This is the right circle of the two circles, because both of the center point coordinates are positive. The x-coordinate of this circle is 2, the radius of this circle is $\sqrt{16} = 4$. To find the value of c, you need to go 4 units over horizontally to the left of $x = 2$. A simple way to write down that last sentence mathematically is:

$$c = 2 - 4 = -2$$

The correct answer is C.

Question 14

In order to perform division of complex numbers you need to multiply the numerator and denominator by the complex conjugate of the denominator. You can get the complex conjugate by changing the sign in front of the complex component of the complex number in the denominator. All of this description of what to do, looks as follows mathematically:

$$\frac{3+21}{2-4i}\left(\frac{2+4i}{2+4i}\right)$$

Perform the multiplication in the numerator first using the F.I.O.L. method. Then combine like terms, keeping in mind that $i^2 = -1$

$$(3+2i)(2+4i) = 6 + 4i + 12i + 8i^2 = 6 + 16i - 8 = -2 + 16i$$

Perform the multiplication in the denominator second using the F.I.O.L. method. Then combine like terms, keeping in mind that $i^2 = -1$

$$(2-4i)(2+4i) = 4 - 8i + 8i - 16i^2 = 4 + 16 = 20$$

After performing this multiplication of the numerator and the denominator, write down the new fraction and separate it.

$$\frac{-2+16i}{20} = \frac{-2}{20} + \frac{16i}{20}$$

Each one of these numbers can have a 2 factored out of it. For clarity factor out a 2 from each number.

$$\frac{-2}{20} + \frac{16i}{20} = \frac{-2(1)}{2(10)} + \frac{2(8)i}{2(10)}$$

Cancel out all the 2s, see how the fractions look and convert them to decimals.

$$\frac{-1}{10} + \frac{8i}{10} = -0.1 + 0.8i$$

The correct answer is D.

Question 15

This question is testing to see if you know the quadratic formula. Before you jump into the quadratic formula however, you need to first rearrange the equation into the appropriate form of a quadratic equation. Do that by adding c to both sides of the equation.

$$-y^2 + dy + c = -c + c$$
$$-y^2 + dy + c = 0$$

This is almost the appropriate form, except that the y^2 term should be positive, if only because it will make life easier for you. To do that multiply the entire equation by -1.

$$-1(-y^2 + dy + c = 0)$$
$$y^2 - dy - c = 0$$

The quadratic formula based off the quadratic equation in the form $ay^2 + by + c = 0$ is:

$$y = \frac{-b \pm \sqrt{b^2 - 4ac}}{2a}$$

For this question the constants a, b and c are as follows:

- $a = 1$

- $b = -d$

- $c = -c$

Plug those values into the quadratic formula to get the following:

$$y = \frac{-(-d) \pm \sqrt{(-d)^2 - 4(1)(-c)}}{2(1)} = \frac{d \pm \sqrt{d^2 + 4c}}{2}$$

The correct answer is B.

Question 16

Set up the equation for the side lengths of a right triangle.

$$a^2 + b^2 = c^2$$

Where the values of a, b and c are as follows:

242

- $a = \overline{AB} = 4$

- $b = \overline{BC} = 3$

- $c = s + 2s + s = 4s$

With those values, the equation for the side lengths of the right triangle looks as follows:

$$4^2 + 3^2 = (4s)^2$$

This can be simplified by squaring the values on the left side of the equation.

$$16 + 9 = (4s)^2 \qquad 25 = (4s)^2$$

Now take the square root of the left and right sides of the equation.

$$\sqrt{25} = \sqrt{(4s)^2}$$
$$5 = 4s$$

To solve for $2s$ divide the equation above by 2, in other words multiply the entire equation by $\frac{1}{2}$.

$$\frac{1}{2}(5 = 4s)$$
$$\frac{5}{2} = \frac{4}{2}s$$
$$2.5 = 2s$$

The correct answer is 2.5.

Question 17

The cosine of a non 90○ angle of a right triangle is equal to the sine of the other non 90○ angle of the same triangle.

The correct answer is 0.7

Question 18

If you replace the a on the right side of the equation with the value of 6, then the right side of the equation becomes $2(6) = 12$. Notice that if you set $x = 0$ then the equation becomes $12 = 12$, which is a true statement.

The correct answer is 0.

Question 19

To solve for the value of q, first solve the lower equation in terms of r. You can do that by first adding r to both sides of the equation.

$$2q - r + r = 3 + r$$
$$2q = 3 + r$$

Now subtract 3 from both sides, so that you can isolate r on the right side of the equation.

$$2q - 3 = 3 - 3 + r$$
$$2q - 3 = r$$

Now plug this value into r in the top equation.

$$7q + 4(2q - 3) = 18$$

Now distribute the 4 over the sum in parenthesis.

$$7q + 4(2q - 3) = 18$$
$$7q + 8q - 12 = 18$$

Next step is add 12 to both sides of the equation and combine the q-terms.

$$7q + 8q - 12 + 12 = 18 + 12$$
$$15q = 30$$

To solve exclusively for q divide the entire equation by 15, in other words multiply the entire equation by $\frac{1}{15}$.

$$\frac{1}{15}(15q = 30)$$
$$\frac{15}{15}q = \frac{30}{15}$$
$$q = 2$$

Take this value for q and plug it into the equation for r.

$$r = 2(2) - 3 = 4 - 3 = 1$$

the correct answer is 1.

Question 20

Add the base salary to the number of washing machines sold multiplied by the amount of money Julie receives per machines sold.

$$100 + (20)13 = 100 + 260 = 360$$

The correct answer is 360.

Question 1

Before trying to solve a cubic equation, simply go through the motions of plugging the possible values into the equation. Start with $t = -3$.

$$(-3)^3 + 2(-3)^2 - 16(-3) - 32 = 7$$

Try the equation for $t = -4$.

$$(-4)^3 + 2(-4)^2 - 16(-4) - 32 = 0$$

The correct answer is B.

Question 2

Jason can dig 300 holes every 8 hours. At an hourly rate that is

$$\frac{300 \text{ holes}}{8 \text{ hours}} = 37.5 \frac{\text{holes}}{\text{hour}}$$

37.5 holes every hour. Jason needs to dig 248 holes, 78 of those holes are already dug. So you are going to need to add the hourly rate Jason can dig holes times the unknown number of hours, h, that he will need to dig to the number of holes already dug and set that sum equal to 248. That will look mathematically as follows:

$$37.5h + 78 = 248$$

The correct answer is C.

Question 3

Separating 30 people into groups one third the size of the entire group will result in groups of 30/3 = 10. Separating those groups of 10 further into fifths will results in groups of 10/5 = 2.

The correct answer is D.

Question 4

If 68.5% of the lawyers prefer using active voice, then 100% - 68.5% = 31.5% of the lawyers prefer using passive voice. Write 31.5% in decimal form as 0.315 and multiply that decimal by the total number of lawyers at 638.

$$638(0.315) = 201$$

The correct answer is A.

Question 5

From the description in the question write down the equation for momentum.

$$P = mv$$

Where:

- P is the momentum

- m is the mass

- v is the velocity

The question is looking for the mass, so rearrange the equation above for m by dividing the entire equation by v, in other words multiply the entire equation by $\frac{1}{v}$.

$$\frac{1}{v}(P = mv)$$
$$\frac{P}{v} = m$$

Now plug in the values of $P = 16$ and $v = 4$ to solve for m.

$$m = \frac{16}{4} = 4$$

The correct answer is C.

Question 6

Because Susy ate 14 more slices of pizza than Jane you can write down the following equation.

$$s = j + 14$$

Where:

- s is the number of slices Susy ate.

- j is the number of slices Jane ate.

Because the total number of slices of pizza eaten by the two is 178 slices you can write the following equation.

$$s + j = 178$$

Solve these set of equation for s by first solving the top equation for j. You can accomplish that by subtracting from both sides of the equation.

$$s - 14 = j + 14 - 14$$
$$s - 14 = j$$

Take this value for j and plug it into the second equation.

$$s + s - 14 = 178$$

Move the constant term to the right side by adding 14 to both sides of the equation.

$$s + s - 14 + 14 = 178 + 14$$
$$2s = 192$$

To solve the equation exclusively for s divide both sides of the equation by 2, in other words multiply the entire equation by $\frac{1}{2}$.

$$\frac{1}{2}(2s = 192)$$
$$\frac{2}{2}s = \frac{192}{2}$$
$$s = 96$$

Now if every pizza has 7 slices, divide the number of slices Susy ate and take the whole number to see how many complete pizzas she ate.

$$\frac{96}{7} = 13.7$$

Susy ate 13 complete pizzas.

The correct answer is D.

Question 7

To find the percentage of coffees sold per day which were mochas, divide the number of mochas sold per day by the entire number of coffees sold per day during the 1990's.

$$\frac{23}{16+2+3+23+22} = 0.35$$

0.35 written in percentage form is 35%.

The correct answer is B.

Question 8

The question says that this line passes through the third and fourth quadrants on the plane. These are the lower left and lower right quadrants. Because the slope of this line is positive, you can imagine the line starting in the lower left corner of the plane and moving up and to the right towards the upper right corner of the plane. If this line has already passed through the lower left and lower right quadrants, it will have no choice but to move on up into the upper right quadrant, which is quadrant 1 or I.

The correct answer is A.

Question 9

Because the question is saying that an animal is being chosen at random, the denominator for this probability calculation is the total number of animals. The numerator will be the total of number of pigs per 100 square miles to be found in Tennessee and Kentucky.

$$\frac{530+413}{14,177} = 0.067$$

The correct answer is D.

Question 10

The fastest race time for a person 72 inches tall can be found as the shortest time run in the points vertically above the number 72 on the graph. Of the three points directly above 72, the quickest race time occurs at the first point above 72, and is 48 minutes.

The correct answer is A.

Question 11

To find the median race time for a person who is 64 inches tall, collect all the race times for the points above 64 and place them in numerical order. That collection of times will look as follows:

$$51, 59, 62$$

The middle number of 59 from this numerically ordered group of numbers will be the median race time of a person who is 64 inches tall.

The correct answer is B.

Question 12

If a polynomial is in factored form such as:

$$P(x) = (x-a)(x+b)$$

The x-intercepts will be at $x = a$ and $x = -b$. Since the question is saying that the intercepts are at 5, 2 and -3, then $P(x)$ must look as follows:

$$P(x) = (x-5)(x-2)(x+3)$$

The correct answer is D.

Question 13

The dollar amount is increasing by a constant rate of 5 every year. Since the rate of change is constant, the dollar amount is increasing linearly over time.

The correct answer is C.

Question 14

If you double x the x becomes $2x$. Plug this into the right side of the equation from the question and see how it changes.

$$\frac{M(2x)^2}{2EI} = \frac{M4x^2}{2EI} = 4\frac{Mx^2}{2EI}$$

Doubling x results in a right side that is four times larger than the right side from the equation in the question.

The correct answer is D.

Question 15

Exponential growth is when something grows at a non-constant rate, in other words, at a rate which continually increases.

The correct answer is A.

Question 16

If they go with company 3, they will accrue the following costs per acre.

- $D = 25$

- $M = 500$

- $L = 400$

Their budget represents the variable I from this question, so that you need to solve the total price equation from the question in terms of A to find the number of acres company X can irrigate with their budget from company 3. To solve for A divide the entire equation by the sum $(D+M+L)$, in other words multiply the entire equation by $\frac{1}{D+M+L}$. This will eliminate the sum in parenthesis on the right side of the equation.

$$\frac{1}{D+M+L}(I=(D+M+L)A)$$
$$\frac{I}{D+M+L}=A$$

Plug in the values discussed for I, D, M and L to solve for the value of A.

$$A = \frac{100,000}{25+500+400} = 108$$

The correct answer is B.

Question 17

The price per acre if Company X were to purchase their irrigation system from Company 1 would be:

$$(17+700+300)A - 20,000 = 1,017A - 20,000$$

The price per acre if Company X were to purchase their irrigation system from Company 2 would be:

$$(13+600+250)A = 863A$$

Set these two prices per acre equal to one another and solve for A. Begin by adding 20,000 to both sides of the equation.

$$1,017A - 20,000 + 20,000 = 863A + 20,000$$
$$1,017A = 863A + 20,000$$

Now subtract $863A$ from both sides of the equation and combine the like terms.

$$1,017A - 863A = 863A - 863A + 20,000$$
$$154A = 20,000$$

To solve for the number of acres divide the entire equation by 154, in other words multiply the entire equation by $\frac{1}{154}$.

$$\frac{1}{154}(154A = 20,000)$$
$$\frac{154}{154}A = \frac{20,000}{154}$$
$$A = 129.8$$

Any acreage below this number, and it would be cheaper to buy the irrigation system from Company 1 with the rebate. Any acreage over this number, and it would be cheaper to go with Company 2.

The correct answer is D.

Question 18

Figure out the volume of each spherical container in gallons. The equation for the volume of sphere is:

$$V = \frac{4}{3}\pi r^3$$

Knowing that the radius is 2 inches you can solve for the volume in cubic inches.

$$V = \frac{4}{3}\pi 2^3 = 33.5 \text{ cubic inches}$$

Convert this value into gallons:

$$33.5 \text{ cubic inches} \frac{1 \text{ gallon}}{231 \text{ cubic inches}} = 0.145 \text{ gallons}$$

Divide the number of gallons of putty by the volume of the spherical containers to find the number of containers she can fill.

$$\frac{14}{0.145} = 96.5$$

She can fill up 96.5 spherical containers and she can fill 96 containers fully.

The correct answer is C.

Question 19

For this question you will need to change the left side of the inequality from $7q + 5$ into $7q - 5$, you can accomplish that by subtracting 10 from both sides of the equation.

$$7q + 5 - 10 \leq 6 - 10$$
$$7q - 5 \leq -4$$

This means that $7q - 5$ can be equal to but not greater than -4.

The correct answer is D.

Question 20

In the first year after the initial population of 10,000, if the population were to triple it would be calculated as follows:

$$P = 10,000(3)$$

For the year after that the population will triple again. Mathematically that will look as follows:

$$P = 10,000(3)(3)$$

So every year you will multiply the previous years population by a factor of 3. The calculation above is after two years, which is $t = 2$. This calculation can also be rewritten as follows:

$$P = 10,000(3)^2$$

Because this is for the second year where $t = 2$, if you were to replace the 2 with a t you will figure out the function to describe the population growth of this city.

$$P = 10{,}000(3)^t$$

The correct answer is C.

Question 21

You are looking for the biggest increase in sales between the two quarters. First look at the black bars and see which one changes positively the most when compared to the grey bar next to it. This occurs when Product C jumps from 6 million in the first quarter to 16 million in the second quarter.

The correct answer is B.

Question 22

If the sales of Product E are growing at a linear rate, then the growth of 6 million sales will be repeated going from the second to the third quarter. Which means that 8 million + 6 million = 14 million sales can be expected in the third quarter.

The correct answer is B.

Question 23

The larger the standard deviation is, the more data points vary largely from the mean. For this mean problem you would expect the mean to be 24 for power supply 1, because you are dealing with a 24-Volt system, and 12 for power supply 2, because you are dealing with a 12-Volt system. If you look at power supply 1 compared to power supply 2 you will notice that very few points are near the mean for power supply 1 and that most of the points are within 1 unit of the mean for power supply 2. This means that the standard deviation is larger for power supply 1.

The correct answer is A.

Question 24

To find the circumference of this cylinder you are going to need to know the radius. Since only the volume of the cylinder is given you are going to need to back out the radius from the volume equation:

$$V = h\pi r^2$$

To solve for r, begin by dividing the entire equation by $h\pi$, in other words multiply the entire equation by $\dfrac{1}{h\pi}$.

$$\frac{1}{h\pi}(V = h\pi r^2)$$
$$\frac{V}{h\pi} = \frac{h\pi}{h\pi}r^2$$
$$\frac{V}{h\pi} = r^2$$

Now to get r all by itself you need to take the square of both sides of the equation:

$$\sqrt{\frac{V}{h\pi}} = \sqrt{r^2}$$
$$\sqrt{\frac{V}{h\pi}} = r$$

At this point plug in the values for V and h, to calculate the value of r.

$$r = \sqrt{\frac{16\pi}{4\pi}} = \sqrt{\frac{16}{4}} = \sqrt{4} = 2$$

The circumference of a circle is calculated with the following equation:

$$c = 2\pi r$$

Where:

- c is the circumference

- r is the radius

Plug the value of $r = 2$ into this equation and solve for c.

$$c = 2\pi 2 = 4\pi$$

The correct answer is D.

Question 25

If you factor out $3t$ from $g(t)$ you will get the following:

$$g(t) = 3t(2t^2 + 5t + 3)$$

Notice that the polynomial in parenthesis is the function $h(t)$. Which means that $g(t) = 3t(h(t))$. Plug this value for $g(t)$ into the solution for $P(t)$ until you can factor $(6t - 4)$.
For solution A, with the substitution for $g(t)$, the function $P(t)$ looks as follows:

$$P(t) = g(t) + h(t) = 3t(h(t)) + h(t)$$

Factor out $h(t)$

$$P(t) = (3t + 1)h(t)$$

This is not divisible by (6t -4)
Notice that as long as the factor in front of $g(t)$ is 1, you will only ever be able to factor out $3t$ from the function $P(t)$. with solution C there is a 2 as a factor for the function $g(t)$. Look at what happens when you plug in $g(t) = 3t(h(t))$ into $P(t)$ for answer C.
For solution C, with the substitution for $g(t)$, the function $P(t)$ looks as follows:

$$P(t) = 2g(t) - 4h(t) = (2)3t(h(t)) - 4h(t) = 6t(h(t)) - 4h(t)$$

Factor out $h(t)$

$$P(t) = (6t - 4)h(t)$$

This is divisible by (6t -4)

The correct answer is C.

Question 26

If you multiply an inequality by a negative number you have to flip the sign around, so that the following is true when you multiply the inequality from the question by negative 1.

$$-1(|a+3| \leq b)$$
$$-|a+3| \geq -b$$

This means that choice I is true.

To remove the absolute value signs, you need to make sure that the value within the absolute value sign is positive. This can be done by setting a limit on a. For example, in this question if you want to remove the absolute value sign you will need to make sure that $a \geq -3$. When you do that you know that $a+3$ will always be 0 or greater. This is the case for choice II, so that choice II is also correct. However, for choice III the same absolute value is removed but the limit on a is set that it is always less than or equal to -3. Doing this will not fulfill the requirements set by the equation $|a+3| \leq b$ in the problem, and only when the inequality sign is flipped around would this inequality be true based on the information given in the question.

The correct answer is B.

Question 27

The price per kilometer traveled, y, is 0.1385 Euros, if you were to travel 0 kilometers. Realistically you would not be traveling 0 kilometers, but it is helpful to imagine that that would be the price if you were to travel 0 kilometers. If you were to travel 1 kilometer you would pay 0.1385 - 1.14×10^{-4} = 0.1384 Euros. This leads to show that 1.14×10^{-4} is the decrease in price for every kilometer traveled.

The correct answer is C.

Question 28

To put this function in terms of the y-intercepts, it needs to be in the following factored form:

$$g(y) = (y-a)(y-b)$$

Where a and b are the negative values of the y-intercepts.
To get the function $g(y)$ from the question into intercept form first rewrite the function as:

$$g(y) = (y+3)(y+3) - 25$$

Now multiply out the sums in parenthesis using the F.I.O.L. method.

$$g(y) = y^2 + 3y + 3y + 9 - 25$$

The next step is to combine the like terms.

$$g(y) = y^2 + 6y - 16$$

Now the time saving trick is to find two numbers which sum up to 6 and multiply to -16. Because the 16 is negative one of the numbers will have to be positive and the other one will have to be negative. The two numbers are 8 and -2, so that the factored form of $g(y)$ looks as follows:

$$g(y) = (y+8)(y-2)$$

The correct answer is B.

Question 29

The variable q is the average of $2a, 3b$ and 56. That can be written mathematically as follows:

$$q = \frac{2a + 3b + 56}{3}$$

The variable r is the average of $4a, 9b$ and 17. That can be written mathematically as follows:

$$r = \frac{4a + 9b + 17}{3}$$

The variable s is the average of $6a, 6b$ and 59. That can be written mathematically as follows:

$$s = \frac{6a + 6b + 59}{3}$$

To find the sum of q, r and s, add up all three numbers.

$$q + r + s = \frac{2a + 3b + 56}{3} + \frac{4a + 9b + 17}{3} + \frac{6a + 6b + 59}{3}$$
$$q + r + s = \frac{12a + 18b + 132}{3}$$

All the numbers in the numerator are divisible by 3. So that you can rewrite the fraction above as:

$$q + r + s = 4a + 6b + 44$$

The correct answer is B.

Question 30

The curve for the function in answer A is a line which passes through the y-axis at the point $y = -2$, it then moves up and to the right, and intersects with the curve shown at two different points.

The curve for the function in answer B is a line which passes through the y-axis at the point $y = -2$, it then moves up and to the right, and intersects with the curve shown at two different points.

The curve for the function in answer C is a horizontal line which passes through the y-axis at the point $y = -2$. It intersects with the curve shown at 4 points.

The curve for the function in answer D is a circle with the equation $x^2 + y^2 = 4$. This circle has its center point at the origin with a radius of 2. It intersects with the curve shown at 6 points.

The correct answer is D.

Question 31

You need to find how long it takes to empty half the gumball machine. First calculate how many gumballs are half of the full gumball machine.

$$\frac{336}{2} = 168$$

If the machine empties at a rate of 12 gumballs per minutes, then you can find out how many minutes it would take to empty half of the machine by dividing 168 by 12.

$$\frac{168}{12} = 14$$

14 minutes is 14(60) = 840 seconds.

The correct answer is 840.

Question 32

To find out how much more gas is needed for each additional passenger change p to $p+1$, that will change the equation into the following:

$$G = \frac{2(p+1)+13}{5}$$

Begin to rearrange this equation by multiplying the 2 over the sum in the parenthesis.

$$G = \frac{2p+2+13}{5}$$

Rearrange the fraction above as follows:

$$G = \frac{2p+13}{5} + \frac{2}{5}$$

The term on the left is the same as G from the question and the term on the right side is the additional gas needed per additional passenger.

The correct answer is $\frac{2}{5}$

Question 33

Call the green beads, g, the orange beads, b, and the total number of beads, t. Because there are 180 more orange beads than green beads, you can write down the following equation.

$$o = g + 180$$

the question also says that $\frac{7}{9}$ of the beads are orange, so that you write down the following equation, which says seven ninths of the total beads are orange.

$$\frac{7}{9}t = o$$

Seven ninths of the total are orange, which means the remaining two ninths must be green.

$$\frac{2}{9}t = g$$

The two equations above are solved for o and g. Plug these values into the top equation, $o = g + 180$, to solve for t.

$$\frac{7}{9}t = \frac{2}{9}t + 180$$

To solve this equation for t, subtract $\frac{2}{9}t$ from both sides of the equation.

$$\frac{7}{9}t - \frac{2}{9}t = \frac{2}{9}t - \frac{2}{9}t + 180$$

$$\frac{7}{9}t - \frac{2}{9}t = 180$$

Combine the t-terms on the left side of the equation.

$$\frac{5}{9}t = 180$$

To get t all by itself multiply the entire equation by $\frac{9}{5}$.

$$\frac{9}{5}\left(\frac{5}{9}t = 180\right)$$

$$t = \frac{9}{5}180$$

$$t = 324$$

Plug this value for t into the equation $\frac{2}{9}t = g$.

$$g = \frac{2}{9}324 = 72$$

The correct answer is 72.

Question 34

For this problem you need to set up some unit conversion ratios and multiply them by 2.5 meters. There are 17 *tacs* in 63.47 centimeters and 100 centimeters in a meter.

$$2.5 \text{ meters}\left(\frac{100 \text{ centimeters}}{1 \text{ meter}}\right)\left(\frac{17 \ tacs}{63.47 \text{ centimeters}}\right) = 67$$

The correct answer is 67.

Question 35

For this problem take d from the equation for I and let it become 2.5d. The right side of the new equation for I, call it I_{new} will look as follows:

$$\frac{\pi(2.5d)^4}{32}$$

Take 2.5 to the fourth power and rewrite the equation.

$$\frac{\pi 39.06d^4}{32}$$

The is equal to 39.06I. Now, the question is looking for the ratio of the original moment of inertia I to the new moment of inertia I_{new}.

$$\frac{I}{I_{new}} = \frac{I}{39.06I} = \frac{1}{39.06}$$

The question is asking you to round to the nearest thousandth.

$$\frac{1}{39.06} = 0.026$$

The correct answer is 0.026.

Question 36

The radius of the unit circle is 1. The circumference is $2\pi r = 2\pi$. three quarters of this value is:

$$\frac{3}{4}2\pi = 4.71$$

The correct answer is 4.71.

Question 37

If the city is losing elevation at a rate of 1%, then the elevation will be 99% of the previous year's elevation in each successive year, 99% is written as 0.99 in decimal form.

The correct answer is 0.99.

Question 38

With the value of $r = 0.99$, set t equal to 8 and solve the equation for E.

$$E = 14(0.99)^8 = 12.9$$

Rounded to the nearest foot 12.9 becomes 13.

The correct answer is 13.

Made in the USA
Middletown, DE
11 February 2020